Franz Hartmann

The principles of astrological geomancy

The art of divining by punctuation

Franz Hartmann

The principles of astrological geomancy
The art of divining by punctuation

ISBN/EAN: 9783742830159

Manufactured in Europe, USA, Canada, Australia, Japa

Cover: Foto ©ninafisch / pixelio.de

Manufactured and distributed by brebook publishing software
(www.brebook.com)

Franz Hartmann

The principles of astrological geomancy

OF

ASTROLOGICAL

GEOMANCY

PUNCTUATION,

ACCORDING TO CORNELIUS AGRIPPA AND OTHERS.

WITH AN APPENDIX CONTAINING 2,048 ANSWERS TO
QUESTIONS.

BY

FRANZ HARTMANN, M.D.

London:

THEOSOPHICAL PUBLISHING COMPANY, LIMITED
7, DUKE STREET, ADELPHI, W.C.

—

1889.

CONTENTS.

———

PREFACE.

THE following book is not intended to be a " fortune teller," but an aid for the student of the higher science, who desires to develop his intuition. A book teaching the art of mixing colours would not make an artist out of a person who has no talent for painting; but to one who is an artist by nature such a book may be very useful. Likewise a work on Geomancy, teaching the rules by which certain truths which are spiritually perceived by the soul, may be brought within the understanding of the external mind, will be of little service to those whose souls have no power of perceiving the truth.

Nevertheless, there is at least a germ of truth in every human being, and every power is developed by exercise. The practice of Geomancy requires above all concentration of thought, and the use of that faculty of the mind by which several ideas may be grasped at once, and be brought to a focus within the field of consciousness, and those who follow the rules prescribed in these pages, may thereby expand their mental faculties and strengthen their intuition, so as to be able to correctly prophecy future events and divine unknown things.

The superficial reasoner, whose mind is captivated by the external and illusive appearance of the world of phenomena, will naturally treat with contempt all knowledge claiming to have been derived from a source superior to reasoning from the base of external observation; but a deeper penetration into the realm of causes will disclose to the lover of truth, the Unity of the divine Law that governs all things, and according to which all external things,

phenomena and events are nothing else but the ultimate
outcome of pre-existing causes in the realm of ideas. The
divine order of things in nature has left nothing to blind
chance; the spirit may know the causes and the mind
calculate the effects. God, the divine self-consciousness in
the universe, knows All, and those who approach in spirit to
the divine source of All will obtain more spiritual knowledge
than those whose minds, wandering away from the centre of
Divine Wisdom, become forgetful of their own spiritual nature,
and lost in the labyrinth of little external details. Those
who desire to practise the art of Geomancy, should
remember that spiritual truth is not found by external
calculation and argumentation, but only by the knowledge
of self.

> "Why idly seek in outward things?
> The answer inner silence brings,
> So to the calmly gathered thought,
> The uttermost of truth is taught."

INTRODUCTION.

The term "divining," or "divination," comes from "*divine*," and the art of divination is based upon the recognition of a universal divine principle acting within the soul. Man is said to be the crown of creation; in him are combined the quintessences of all the four kingdoms. He is at once a mineral, a plant, an animal, and a god, and each of these constituent parts has its own peculiar states of consciousness, its own sensations, desires, feelings, and perceptions. The divine Light which shines within the darkness of his material constitution is the eternal Spirit of God, in which there exists neither past nor the future time, but in whose consciousness all things are forever present. Its presence is felt within the soul as the divine power of Intuition, and if the Mind of man were to rise entirely above the realm of selfishness, to become illuminated by the Light of the Spirit, there would be no need for Geomancy or for any other artificial aids to bring the knowledge of the Spirit to the understanding of the material Intellect. We could then not only intuitively *feel* the truth, but *see* it and *know* it without any argumentation or mathematical reasoning.

There are, however, only a few saints or adepts in the world who are in possession of such a state of perfection, and the majority of men and women upon this globe have to go the roundabout way of speculation and calculation to obtain information in regard to things unknown. The psychological process by which the knowledge of the spiritual soul comes to the understanding of the human intellect appears to be very complicated; it seems that the divine ray of Light has to pass through many *strata* of matter, and is broken many times, before it is ultimately reflected within the field of external consciousness, and the more we are able to spiritually rise above these clouds of matter that darken the mental sky, the more will we become able to see the sunlight of truth in its purity.

By practising the art of Geomancy in that state of mind and feeling which brings Man nearer to the perception of the Truth, the Intuition may teach the reasoning Intellect. The first four symbols of which the geomantic figure is constructed, are the products of Intuition, and from them the final result is obtained by intellectual labour. The condition for obtaining success is the full and entire concentration of thought and will upon the question which is to be answered.

In the art of Geomancy it is not the mind, but the soul which answers the question and the answer is received by means of the power of the living divine Spirit of God, whose temple is man. It is, therefore, clear that this magical art ought not to be practised in any other frame of mind than that of worship, adoration, and faith in the eternal Law of order and harmony. If undertaken merely for the purposes of gratifying idle curiosity, or for selfish purposes, or from motives of greed or revenge, its results will be unreliable; because in such cases the intuitional ray becomes distorted by the perverted images existing within the mind. Likewise the answers will only be reliable, if the whole strength of will and thought is concentrated upon the question asked; a vacillating mind has but little power, for the truth suffers no other loves, only those who rise up to it and embrace it with all their soul, with all their mind, and with their whole being, will receive true knowledge.

Finally, it may be well to state that the foreseeing or future events will not enable us to change their course; for if these events did not already exist in the future, they could not be foreseen, while if they were to be altered, the alteration would also be seen. Nevertheless, the art of

Geomancy may be very useful to act as our guide in meeting future events, in enabling us to take courage in what we know to become a success, and to give up insisting in fruitless attempts to attain that which would end in a failure if carried out. It should, however, always be kept in mind that the reliability of all such arts depends on our own reliability of perception, and the decisions of Geomancy can be infallible only when all the conditions required to make them infallible have been complied with. Geomancy is not a substitute but an aid for divine Reason.

PLANETARY SIGNS.

☉ The Sun.	♀ Venus.	♄ Saturn.
☽ The Moon.	♂ Mars.	♅ Uranus.
☿ Mercury.	♃ Jupiter.	♆ Neptune.

—

SIGNS OF THE ZODIAC.

NORTHERN.	SOUTHERN.
♈ Aries.	♎ Libra.
♉ Taurus.	♏ Scorpio.
♊ Gemini.	♐ Sagittarius.
♋ Cancer.	♑ Capricornus.
♌ Leo.	♒ Aquarius.
♍ Virgo.	♓ Pisces.

ASTROLOGY.

THE science of Astrology is based upon a correct understanding of the true nature of Man and his position in the Universe. Natural man is not, as many vainly imagine, a self-existent being, creating his own ideas, thoughts and feelings; but as his physical body is the product of the confluence and assimilation of physical atoms, likewise the constitution of his mind is the product of the action of the intellectual and emotional elements entering his psychic organism. His visible and tangible form receives from the great storehouse of physical nature the four originally invisible elements, called : Earth, Water, Fire and Air (solid, fluid and gaseous essences, heat, electricity, life, etc.), and by means of the physiological processes going on in his visible form he transforms them into such substances and activities as are required by the nature of his corporeal body. These processes are taking place without the intellectual supervision of man ; they are going on instinctively, involuntarily and even without his being conscious of it.

All this nobody will deny ; because we see the food which we eat and the water we drink ; we know of the existence of the air which we inhale, and we feel the heat that warms our bodies. These things are not our own creations, nature prepares them for us, lends them to us, and after we have made use of what we have borrowed, we restore them back to nature.

With the invisible principles that enter the invisible soul of Man, the same process takes place. We do not create our own thoughts, but the ideas which in pure, permanent and indivisible incorporeal ideal forms exist in the *Astral*

Light, reflect their images in the individual minds of men and women, in the same way as a landscape may be reflected in a looking-glass, or the whole of the starry sky be mirrored forth in a drop of pure water. These images may enter the consciousness of man without any voluntary effort on his part, in the same sense as the air enters his lungs without man's conscious effort to breathe. They may even enter against his will and desire; for there are unwelcome thoughts that come when they are not wanted, and there are welcome ones difficult to retain. Animals only think the thoughts which enter into their minds without any efforts on their part; but man has the power to voluntarily rise in thought to the realm of ideas and grasp the images which he desires, and therefore it is said that man does not need to be governed by the stars, but that he can become superior to them.

The ideas which enter the field of consciousness act upon his Imagination, and his Imagination re-acts upon his Will; thereby producing certain states of feelings or emotions according to the nature of the idea, from the most gross and vulgar passion up to the highest state of exalted thought. As the food which he eats determines the state of purity or impurity of his physical organism, so likewise the thoughts which he harbours, and the feelings in which he indulges, determine the purity or impurity of his soul.

Man does not create his thoughts; but he elaborates them from the ideas which he absorbs, in the same sense as his physical body elaborates the food which he eats, and transforms vegetables and grain into blood and flesh and bones. Likewise the mind of man mixes and combines ideas, and infuses then with life by the power of his Will; and as an impure body may poison the surrounding atmosphere by the unhealthy effluvia coming from his impurities, so also the emanations of an impure mind will poison the mental atmosphere with the products of an impure Imagination rendered alive by an evil Will; for the products of thought are rea and substantial things, even if they are invisible for gross material eyes, and the Will is a real power which may act as far as thoughts travel.

As the moon without the light of the sun is dark, so likewise the images produced by thought have no power unless they are strengthened by the Will; while the Will is useless unless it is guided and brought into a form by thought. If thought and will are divided, they are both ineffective; but if thought and will are in unison, they become effective; they then constitute a Unity, and this unity is called "*Spirit.*"

According to the statements of the wise, all things in the universe are the products of Will and Imagination (Ideation) acting in unison, and, therefore, all things are produced by Spirit. Spirit is the Reality; that which we call the material form is merely the shadow of the light of the divine Ideal. That which we call "Matter" is the same thing as "Spirit," only it is in a state of inertness or condensation, while the vibrations of Spirit are far higher than those of Matter, so as to transcend our physical power to perceive them, in the same sense as there are vibrations of sound too high to be heard by the external ear, and vibrations of colour too high to be seen by the external eye. We are spirits ourselves, even if we are clothed in a gross material form; we live in a spiritual world; we are surrounded and permeated by Spirit. We are continually influenced by spiritual powers that come to us with or without our desire, and we have no other means to protect ourselves except Reason and Will. Man is a little world in which all the celestial and terrestrial powers and principles existing in the great universe may be reflected, and upon the perception of this truth rests the science of Astrology.

Everywhere in the universe rules the same fundamental law; everything is governed by order and harmony. The planets revolve in their orbits with mathematical precision, and each human being must follow the orbit in which it has to revolve. A man may oscillate to the right or the left as

he traverses his orbit, but he, cannot leave the line of his
destiny, which is the result of causes produced in previous
states of existence. There are ebbs and tides of the sea, and
there are ebbs and tides in the ocean of thought. There are
conjunctions and oppositions of spiritual influences in the
world of ideas, as there are among the corporeal planets.
There are times in which mankind as a whole rises up to a
higher state of spiritual enlightenment, coming nearer to
God ; and there are other times in which they sink deeper
into ignorance and superstition.

If the mathematical rules that govern in the realm of ideas
were as well known as those that rule the revolutions of the
visible planets, there would be as little difficulty in predicting.
future thoughts and the external events resulting therefrom,
as there is in predicting an eclipse of the moon ; but as long
as the mind is too much captivated by the external sensual
impressions to be attracted by the things that belong to the
Spirit of Truth, its deductions and conclusions will be unreliable. As long as its imagination is captivated and its
desire inflamed by the fire of the Astral plane, the thoughts
and aspirations will not penetrate into the pure region of
Truth.

Fortunately, however, there is a knowledge higher than
that of the speculating brain ; it is the knowledge of the soul,
and if there were nothing to prevent the free communication
of the Intellect with the Soul, man might intuitively know
many things which now seem beyond the reach of his
knowledge.

But it is written, that "those who desire to live from the
altar must serve the altar," that means to say, that those who
wish to know and be rendered alive by the truth must serve
the truth by loving it with their whole heart, and manifest
that love in thoughts, words, and actions. Those who desire
spiritual knowledge, and to obtain the power to predict
future events, must above all seek for the truth within their
own souls. They should put away their passions and evil

desires, their scientific, social, and religious prejudices, and the errors which have been engrafted into their minds by a false education, received in an age of so-called rationalism, during which there was but little comprehension of spiritual things; when the Sun of Divine Wisdom was obscured by the delusive shadow of the speculative semi-animal intellect, and when the voice of the Intuition was drowned in the noise made by the clamours of conceited ignorance assuming the place of science.

Fortunately the days of *Kakosophia* are approaching their end, and humanity as a whole is again coming nearer to the zenith of ☉. May all the lovers of truth make use of this opportunity to enjoy all the light which they are capable to receive, before the planet, following the law of order, will again descend into the shadow of ☾.

The principle upon which astrology is based, cannot be fully understood, unless the nature of the "planets" with which it deals is realized; but this realization will not be the product of book reading, nor can a person give to himself a knowledge or power which he does not possess; it will remain unattainable to the doubter, and can only be acquired by *faith*; that is to say, by the *interior awakening* of the *Spirit of Truth*.

The Seven Planets.

THE ancients recognized the presence of seven different states of the One Universal Spirit which constitutes the Soul of all things as well as all physical bodies, and they gave to these seven principles the following names and symbols, which are also those of the "seven planets" and of the seven days of the week—

♄ Saturn, Saturday.
☉ Sun, Sunday.

☽ Moon, Monday.
♂ Mars, Tuesday.
☿ Mercury, Wednesday.
♃ Jupiter, Thursday.
♀ Venus, Friday.

It is almost unnecessary to say that these " seven planets " have little or nothing to do with the seven cosmic bodies in our solar system bearing the same names ; for although the latter may to a certain extent be regarded as their external and visible representatives, the principles themselves are invisible, and rule not only within our solar system, and throughout the extent of the Macrocosmos of the Universe ; but also within the internal constitution of the Microcosm of Man. Their significations differ according to the aspects which we take of them. Generally speaking, they may be stated as follows :

The Sun is the emblem of Wisdom. In him are the powers of all the planets united; in him are love, will, and intelligence combined into one; in the same sense as the four sides of a pyramid all culminate in one point. The Sun is the centre and source of all light and heat, and of all power ; not only of the visible terrestrial light, but of the light of intelligence; not only of terrestrial heat, but of the heat of love. He attracts by his power all the planets in space and keeps them within their orbits. Those in whom the sun principle is strong are capable of becoming wise, strong, and powerful. It is therefore said that the Sun is a planet governing the souls of kings and noblemen, and conferring honours, powers, and titles. Its influence is decisive in all important questions in human life. In the mineral kingdom it is represented by gold ; in the animal kingdom by the Lion, in the spiritual kingdom as *Sol-om-on*, the divine Sun of Wisdom.

The Moon is the symbol of imagination, illusion, and dreams. She has no light of her own, but borrows her light from the Sun. Without the light of the sun the moon would be cold and dark; without the power of the Will the products of the Imagination are without life. Thoughts become powerful only when they are infused by the will; they become luminous only when they are illuminated by love; they can be wise only if permeated by wisdom. Under the influence of the moon are said to be especially dreamers and mediums, persons who live a great deal in the realm of imagination and fancy, ladies of rank, pleasure seekers and travellers; it is said to govern things in which there is little firmness and stability, especially water and ships. In the mineral kingdom the Moon is represented by silver, in the spiritual kingdom by *Luna*, the queen of the night.

Mars represents strength. If unguided by wisdom it is a dangerous planet, inclined to deeds of violence, acting rashly and without consideration. It is a principle which causes anger and wrath. It has been regarded as the god of warriors, soldiers, lawyers, causing also the effects of violent medicines. Its action may become moderated by its union with ♀. Among the metals Mars is represented by iron, in the kingdom of spiritual powers by the god of war. It is of a fiery nature, and as fire does not combine with water, likewise an irate temper and the assertion of self-will is incompatible with that calm and peaceful thought necessary for the perception of the truth.

$$\underset{\Large\Phi}{\text{☿}}$$

Mercury represents the Intellect, and it may be a good or evil planet according to the conditions under which it acts. If ☿ is under the influence of ♄ ; that is to say, if the intellect is subservient to selfish and material things, it becomes a source of evil; if combined with ♃ it will produce pride ; if united with love ☿ it will become wise, and in this manner " crude mercury " may be transformed into the gold of wisdom.

Mercury without love is said to rule especially those who live by their wits; scientific speculators, sophists, merchants, thieves, intellectual but not necessarily moral persons, men of letters, students, etc. In the mineral kingdom it is represented by quicksilver, in the spiritual realm by the god of trade.

$$\text{♃}$$

Jupiter represents power. Its qualities differ according to ts aspects. Its symbol is an eagle ; because it enables man to rise up by its power into the highest regions of thought, even to the throne of the Eternal. It is, or ought to be, therefore, the ruling planet for ecclesiastics and clergymen, and those who have to deal with the administration of justice. Its influence gives eloquence. It is friendly with all the rest of the planets except ♂; the latter being loved by none except ♀. In the mineral kingdom it is represented by tin; in the spiritual realm by Jupiter, the king of the gods, who obtain their power through him.

♀

Venus represents love. In its lowest state it is blind attraction, producing gravitation among the corporeal planets and instincts among animals. The more it becomes amalgamated with intelligence, the more does it become capable to manifest its divine qualities. Pure love is a divine and self-existent power which only gives and does not seek to receive. It has no desires; but it creates desires in the objects in which its power awakens. In its higher aspects it rules artists and true physicians, in its lower state it is active in all affairs of love and marriage and in pleasures of various kinds. Among the metals it is represented by silver, in the spiritual realm by the goddess of Love. As Power is the father of all the gods, likewise Venus is their mother. No being can exist without love. When their ♀ is departed they will all be swallowed up by ♄.

♄

Saturn represents the element of Matter. Not the visible tangible earth, but the primordial Substance out of which all things are made. It is also the principle of Life. It produces and destroys all forms, and is therefore represented as the god who eats his own children. Unless associated with ☉, Saturn is a cold, cruel and dark planet. It therefore rules old persons, misers, and usurers, gross material and vulgar people, and governs agricultural and mining pursuits. In the mineral kingdom ♄ is represented by lead; in the spiritual realm as the god of Time.

Saturn represents darkness and fear, melancholy, gloom, and death; but it is also the god of Life, for all so-called death is merely a change of state, and in the end of an old form, is the beginning of a new state of being.

CONJUNCTIONS.

From the approximations and conjunctions of planets, or, in other words, from the combination of different emotions and mental states, result a great many varieties of powers, which again differ from each other in regard to their quality of gradation.

In every kingdom of nature we find sympathics and antipathics. Among the planetary influences they are as follows :—

☉	sympathetic to	♃, ♀	antipathetic to	♂, ☿, ☽
☽	,,	♃, ♀, ♄	,,	♂, ☿
♂	,,	♀	,,	☉, ☽, ☿, ♃, ♄
☿	,,	♃, ♀, ♄	,,	☉, ☽, ♂
♃	,,	☉, ☽, ☿, ♀, ♄	,,	♂
♀	,,	☉, ☽, ♂, ☿, ♃	,,	♄
♄	,,	☉, ☽, ☿, ♃	,,	♂, ♀

The Colours of the Planets :—

☉ yellow.	☽ white.	♂ red.	☿ brown.
♃ blue.	♀ green.	♄ grey.	

One colour may be changed into another by mixing it with one of a different kind; for instance, blue mixed with yellow will produce green, and likewise a planetary influence may be changed in its nature by coming in conjunction with one of another kind. Love and Imagination mixed together may in their exultation result in insanity, ♂ and ☽ produce hallucinations, ♃ and ☽ will make a man very vain, ♂ and ☿ may make a robber out of a thief, ♃ and ♀ may be the source of the inspiration of an orator, and if ☿ is added, carry him up into the highest regions of thought. In this manner an almost endless number of combinations may be formed in the alchemical laboratory of the soul. There is no action without reaction. Each planet has therefore a

two-fold aspect, and may manifest a two-fold activity.
There is no relative good without relative evil, and no evil
without good. The same ☉ that is the source of all life may
burn up living forms, ♀ coming in opposition to ☿ may turn
into hate; the same ♃ that lifts the aspiring mind up to
heaven may, by becoming perverted, turn into an angel of
evil and drag him down into the abyss of self-conceit.

Thus the lover of astrology will find abundant material in
his own mind to study the various states resulting from the
approximation, conjunction, or opposition of planets, without the aid of books, and such a study will be found to be
even more interesting and useful in the end than that of the
stars in the sky, for however sublime the study of the
latter may be, all the learning of external things in the world
is far inferior to the knowledge of self.

THE TWELVE SIGNS OF THE ZODIAC.

The twelve signs of the zodiac, or the circle through which
the Earth travels in its annual revolution around the sun, are
described in every almanac; but in their deeper signification they represent principles which form the basis of the
evolution and involution of the Universe. Their names and
signs are as follows:

△ ♈ Aries.
+ ♉ Taurus.
= ♊ Gemini. } Ascending signs.
▽ ♋ Cancer.
△ ♌ Leo.
+ ♍ Virgo.
The point of equilibrium= ♎ Libra.

♏ Scorpio.
♐ Sagittarius.
Descending signs { ♑ Capricornus.
♒ Aquarius.
♓ Pisces.

The twelve signs of the zodiac represent the powers of Man; but as these powers are of a spiritual nature, their qualities can be known only to those who in the course of evolution and unfoldment have become conscious of their existence. The more we enter the realm of spiritual know-ledge, the more do we begin to realize that all terrestrial knowledge is but child's play in comparison with the true knowledge of the spirit; but the door to the temple in which the truth can be seen without a veil is guarded by the dragon of selfishness, and only those who are able to conquer the " beast" can enter the sanctuary.

No greater truth has ever been pronounced than that we must die before we can begin to live. The mysteries of the inner temple can never be divulged to the uninitiated, because they would not be understood even if an explanation were attempted. Sensual things may be perceived by the senses; intellectual verities may be intellectually under-stood; but only the Spirit of God in Man searches the depths of Divinity. Sciences may be taught, arts may be acquired by practice; but divine Wisdom can only be attained by the grace of the divine spirit, and all that man can possibly do is to render himself capable for its reception, by seeking to eliminate from his constitution those elements which hinder the entrance of Light. Therefore the religious books advise us that man should seek above all the Kingdom of God (the divine consciousness), promising that then all other kinds of knowledge will be given to him; but of the scoffer and sceptic who desires to prey with curious eyes behind the veil it is said that his safety is in his ignorance, for the knowledge which he might attain would prove to be the cause of his perdition.

To give a correct and complete description of the twelve signs of the zodiac would require the writing of a book greater than all the books in the world, nor could any amount of words be adequate to describe the sublimity and grandeur of thought and conception necessary to rise up to a comprehension of one of the greatest of all divine mysteries, the construction of the spiritual and material universe, or in other words, of Nature in her aspect as the living temple of God.

If we therefore attempt to describe our ideas in regard to the character of these divine principles, we are well aware of the difficulty of our task, and it must be left to the reader to ask for more light, by seeking within himself for the hidden truth.

♈

Aries, or Ram, represents the universal principle of Life or ☉, which is the source of all things. It may also be represented as ♄, or the universal element of living Matter or Substance, in which all things are, and by whose power they all exist. It is " Matter " and " Force " in one ; for those terms do not represent two things essentially different from each other ; they are merely two words representing two states of the eternal *One*, for which there is no name. " Matter " means relative inactivity; " Force " means a higher state of activity of the same principle. The spirit of man descends into " Matter," that is to say, it becomes relatively inactive and unconscious, and re-ascends again to its highest state as a self-conscious spiritual power. The process taking place in each individual monad corresponds to the grand process taking place in the evolution and involution of the universe as a whole.

♉

Taurus, or the Steer, represents Power. It symbolises the divine power of that universal principle, which is at once the creator, preserver, and destroyer of forms. By the inherent strength of the divine principle in Man, humanity is enabled to aspire for something higher than merely

material existence, and to rise up to its former divine state
as a spiritual self-conscious being.   In one of its aspects, ♉
may, therefore, be compared to ☾; because the light of the
Spirit begins to be reflected by the material mind.   In
another aspect it resembles ♀, for all power originally
arises from Love, and in still another it may be compared
to ♃, for in this sign Man begins to realize the glory of God.
In fact, each of the zodiacal signs may be compared to all
the planets; for Spirit is a Unity, and in each sign are,
therefore, contained the powers of the other six.   The
distinction is not made on account of any difference in their
essential nature; but in regard to the form of their
manifestation.

♊

Gemini, or the Twins, represent the spiritual Man of whom
the mortal body is merely an imperfect image or a reflection.
It is the personal God of each man, the divine *Adonai* who is
neither male nor female but in whom both sexes are united
by the divine marriage of Intelligence and Love.   It may be
said to correspond to ☿ united with ♀, or to a union of
Will and Thought.   Its germs are within every man or
woman, for in each human being are male and female elements; a being wholly male would have no Will, a being
wholly female would have no Imagination.   In its aspect as
a universal power ♊ represents the "Great Spirit," the universal bisexual *Man*.

♋

Cancer, the Crab, represents retrogression; that is to say,
the final descent of the Spirit from its divine into a material
state by the act of creation.   It also represents the power of
the "*Word*," through whose action creation takes place.   It
is the Λ and Ω, the beginning and the end; if the word had
never been spoken, there would never have been any objec-

tive creation and God (resp. *Man*) would not have left his divine state of rest. Perhaps it may be compared to ♂, as it also represents Power acting regardless of personal danger, and without any consideration of the consequences that must result to itself from its entering the realm of darkness, the material plane; impelled as it is by the power of ♀; for thus has ☿ loved the world, that it sent its own essence and the power to enter the hearts of mankind, and to redeem them from the realm of illusions.

♌

Leo, Lion, represents that divine power of Christ, the Man anointed with spiritual knowledge, which enables him to rise again up to a conception of his own divine state. It may be compared to ♃, or the *Eagle*, who wings itself up to the throne of the Most High. It represents the true saviour of mankind; for no one can enter the kingdom of heaven unless he has the power to do so, and no one can come to the unmanifested God except through the power of the Christ in whom God has become manifest. Let those who desire to enter the kingdom by force, meditate about the signification of ♌, and enter into this sign; for the kingdom of heaven can be gained only by violence.

♍

Virgo, the Virgin, represents the Spiritual Soul in Man and in the Universe, the celestial virgin, the eternal mother of man-made God. It may be compared to ☾, in which the power of ☉ is reflected, and becomes substantial light; it is divine Intuition, which saves from perdition the semi-animal Intellect; it is for ever immaculate, because it has nothing to do with external reasoning and argumentation; it knows the Truth, because it is one with it. It represents

Isis, the eternal goddess of Nature, from whose womb will arise *Horus*, the time-born god. It is the eternal patroness of those who seek for salvation, as its exalting influence raises man up into a higher region of thought. It is one of the greatest mysteries of religion, and can form no object for external scientific research.

───⌒───

Libra, the Balance, represents the point of equilibrium, the unimaginable state of *Nirvana*, which cannot be described, and for which we have no words in the language of mortals to explain its condition. If a planetary sign is to be compared with it, it must be ♀, for Love is the root out of which all powers spring.

When the Day of Creation* is over, the celestial powers again retire to the bosom of their eternal Father, to rest in divine blissfulness until the equilibrium becomes again disturbed by the awakening of an internal desire for creation and a new evolution begins. The descent of Spirit into matter then recommences, producing once more the "fall of man," and this work of involution is represented in the following descending signs of the Zodiac.

♏

Scorpio, or Scorpion, represents that *Desire* for knowledge which again induces the celestial spirit to descend and to overshadow material forms. It is the *Snake* which eternally tempts *Eve* to break the fruit from the forbidden tree and to offer it to the Intellect for its comprehension. By its influence the attention of the Spirit of Man is again attracted to the realm of phenomena, and he again enters the wheel of evolution, but on a higher scale than before. In its

---

* We purposely use the term "creator" for the constructor of forms; because forms are nothing, they are merely appearances, and in creating a form an illusion is *created*.

universal aspect it represents that state of the Universal
Mind in which the idea of a new creation begins to exist.
If we had to compare it with any planetary sign we would
choose for it ♂, or Love acted upon by the Will.

Sagittarius, the Archer, represents the divine Will to create
a new world, for the thought alone would not be sufficient to
produce a world existing in the imagination, unless the
divine will were present to project it into objectivity. In
a lower aspect it may represent that power by which the individual spirit, unable to create a world of its own, is impelled
to form again a connection with Matter. In its aspect as a
universal principle it may be compared to ♂.

Capricornus, or the Goat, represents the exercise of the con-
structive power of the universe: the universal Law of
Evolution, which at the time of the beginning of a new
creation again enters into activity. It also is the symbol of
perversion, and in its lower aspect it may signify the power
by which the disembodied spirit again builds up a mortal
house of clay, and enters again into the world of formation.

Aquarius, the Waterman. The product of the imagination
acting within the Will is Thought. Water is the symbol
for Thought, and a " Waterman" therefore symbolizes a

man formed of thought. All things are made of thought ;
the visible as well as the invisible world is the product of
Thought and made up of the substance of Mind ; material
forms are merely the external expressions of internal
principles which must necessarily be substantial, for forces
are states of matter, and a state of a thing which does not
exist is an imposssibility. The whole universe is substantial thought, produced by the will and the imagination. ♒
may therefore be regarded as the creative power of divine
Will and Thought; that is to say, as the *Word*, in its aspect
as a universal divine Power. In its more limited aspect, it
represents the power by which the Spirit again assumes a
material form.

♓

Pisces, Fishes. The fish lives in water; man is living in
the ocean of thought, rendered more or less material by the
influence of ♄ . In one of its aspects, ♓ represents man
as a being immersed in an ocean of spiritual ideas, and therefore those who are supposed to live in that higher region of
spiritual and exalted thought have been called episcopes or
or bishops. In another aspect this sign may represent the
world of ideas existing in the Astral Light. Gradually the
wheel of evolution approaches again the sign of ♈, representing the realm of Matter ; the descent of the spirit is then
accomplished, and the ascension begins, unless man should
allow himself to descend still deeper into the realm of darkness where those who wilfully reject the light of the Spirit
are doomed to perish. When the descending Spirit in whose
idea the world of formation exists, expresses his thought in
the act of creating, he finds himself like a fish in the water,
in the world of forms which he has created himself, and
then the reign of ♄ begins again, and with it the work
of redemption by the ascending signs,

THE SYMBOLS OF GEOMANCY.

There are sixteen geomantic symbols, corresponding to certain planetary signs, as follows :—

Fortuna major.

.

Fortuna minor.

Via.

Populus.

Acquisitio.

Lætitia.

Puella.

Amissio.

Conjunctio.

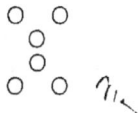

Albus.

Puer.

Rubeus.

Carcer.

Tristitia.

Caput Draconis.

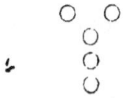

Cauda Draconis.

SIGNIFICATIONS.

Via, Street or Way, is neither good nor bad; its quality, like those of all the rest, is determined by its position in the house of the astrological figure. Its nature corresponds to ☽, its element is watery, its zodiacal sign ♌, and its number 7.

Populus, People, is likewise indifferent. Its nature corresponds to ●, its element is watery, its zodiacal sign ♑, and its number 16.

Conjunctio, union or coming together, is rather good than bad. Its nature corresponds to ☿, its element is airy, its zodiacal sign ♍, and its number 11.

Carcer, prison, or to be bound, is good or bad, according to the nature of the question. Its nature corresponds to ♄, its element is earthy, its zodiacal sign ♓, and its number 10.

Fortuna major, Great Fortune, success, interior aid and protection, is a very good sign. Its nature corresponds to ☉, its element is earthy, its zodiacal sign ♒, its number 12.

Fortuna minor, Little Fortune, external aid and protection, is not a very good figure. Its nature corresponds to ☉, its element is fiery, its zodiacal sign ♉, and its number 10.

Acquisitio, Success, obtaining, absorbing, receiving, is a good figure. Its nature corresponds to ♃, its element is airy, its zodiacal sign ♈, and it number 7.

Amissio, Loss, that which is taken away. It is a bad figure. It nature corresponds to ♀, its element is fiery, its zodiacal sign ♎, and its number 8.

Laetitia, Joy, health, laughing, is good. Its nature corresponds to ♃, its element is airy, its zodiacal sign ♉, and its number 15.

Tristitia, Sorrow, grief, perversion, condemnation, is bad. Its nature corresponds to ♄, its element is earthy, its zodiacal sign ♒, and its number 14.

Puella, Girl, pretty face, is pleasant but not very fortunate. Its nature corresponds to ♀, its element is watery, its zodiacal sign ♎, and its number 2.

Puer, Boy, rash and inconsiderate, is rather good than bad. Its nature corresponds to ♂, its element is fiery, its zodiacal sign ♈, and its number 3.

Albus, White Head, wisdom, sagacity, clear thought, is a good figure. Its nature corresponds to ☿, its element is watery, its zodiacal sign ♋, and its number 12.

Rubeus, Redhead, passion, vice, fiery temper, is a bad figure. Its nature corresponds to ♂, its element is fiery, its zodiacal sign ♊, and its number 13.

Caput Draconis, Dragon's head, entrance, threshold, upper kingdom, is good. Its symbol is ☊, its zodiacal sign ♍, its element earthy, and its number 4.

Cauda Draconis, Dragon's tail, exit, lower kingdom, is bad. Its symbol is ☋, its zodiacal sign ♐, its element fiery, and its number 5.

The significations of these symbols differ to a certain extent according to the nature of their origin.

A good figure made of two good ones is good.

A bad figure made of two bad ones is bad.

A good figure made of one good and one bad figure means success, but delay and vexation.

If the two witnesses are good and the judge bad, the result will be obtained; but it will be unfortunate in the end.

If the first witness is good and the second bad, the success will be very doubtful.

If the first witness is bad and the second one good, the unfortunate beginning will take a good turn.

C

INSTRUCTIONS FOR THE PRACTICE OF GEOMANCY.

PREPARATION.

THE art of Geomancy ought not to be practised unless the mind is tranquil and calm. If the field of mental vision is clouded by fear or doubt, by grief or selfish desires, if the temple of the spirit is occupied by money changers or resounding from the quarrels of the pharisees and scribes, it will be difficult to hear the voice of the truth. Cornelius Agrippa says that Geomancy should not be practised " on a cloudy or rainy day, or when the weather is stormy, nor while the mind is disturbed by anger or oppressed with cares." Neither should it be practised for the purpose of gratifying idle curiosity, for mere amusement, or for giving tests to the sceptics. Finally, the same question ought not to be asked repeatedly in the same form.

It is, furthermore, desirable that for each question an appropriate day and hour should be selected ; for instance, all questions in regard to agriculture or mines should be asked on the day, and in the hour of Saturn, all questions in regard to love and marriage on the day, and in the hour of Venus, etc.

The reason for this is that the soul of man stands in intimate relationship to the soul of the world, and that the influences of the higher world act correspondingly upon the lower. If man were in a perfectly natural state, and his soul in exact harmony with nature, he would be more sensitive to planetary influences, and the state of his thoughts and feelings would correspond to the states of the Universal Mind.

The days of the week are named by their ruling planets, and each planet rules the first hour of his day. To find the planetary hour, it is, therefore, merely necessary to divide the time from sunrise to sunset into twelve equal parts, adding, however, one hour for the twilight. The first planetary hour is then dedicated to the planet of the day, and then follow the other planets in their regular order : ♄ , ☉, ☽, ♂ , ☿ , ♃ , ♀ , ♄ .

Thus, if we imagine some Wednesday in summer, when the sun rises at five and sets at seven, we will have a day of fourteen hours, to which one hour is added. These fifteen hours are divided into twelve equal parts. The hour of ☿ then begins at 4.30 A.M. and ends at 5.45. Then follows ♃ ending at 6.60 ; next ♀ , etc.

However useful it may be to observe the planetary hours, it is undoubtedly of still greater importance to pay attention to the constellation within, and to take care that no evil spirit enters the sphere of mind of the operator to interfere with his work.

The Practice of Geomancy.

If it is desired to obtain by the art of Geomancy an answer to a certain question, it is, above all, necessary to be in a tranquil state of mind, and to fix one's thought firmly, and without wavering, upon that question. While the mind is thus fixed, the right hand is used to make an indefinite number of points without counting them from right to left in the following manner :

.   .   .   .

They may be made with a pencil upon a piece of paper, or following the old custom, with a stick on the ground, and it is from this old method that the term " Geomancy " (Divining by means of the earth) is derived; for it is believed by some that the Elemental spirits of Earth are guiding the hand of the operator. Four such lines are constructed, and each line will have either an even or an uneven number of dots.

If the number is even, two marks are made at the end of the
line; if it is uneven, only one. The four lines produce one
geomantic figure, as may be seen in the following example:

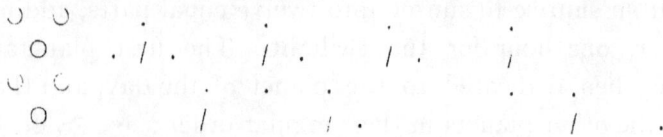

This process is repeated three times.

In this way four figures have been obtained; they are
called the *Mothers*.

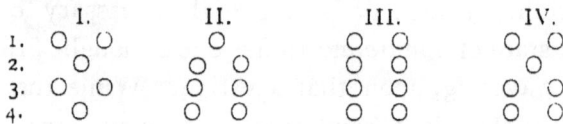

Each of these figures consists of four parts: 1, 2, 3, and 4.
1 is called the heads; 2, the necks; 3, the bodies; and 4, the
feet. By taking the heads of the four mothers, and putting
them below each other, the first daughter is produced; the
necks produce the second, the bodies the third, and the feet
the fourth daughter, as follows:—

DAUGHTERS.

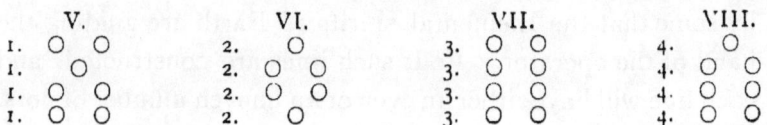

The Nephews are produced in a different manner. To produce the first nephew the heads of I. and II. are counted together and marked down as even or uneven, then the necks, next the bodies and the feet.

The second nephew is produced in the same manner from III. and IV., the third from V. and VI., the fourth from VII. and VIII.

NEPHEWS.

IX.	X.	XI.	XII.
O	O O	O	O
O	O	O	O O
O O	O O	O O	O O
O	O O	O	O O

From the four Nephews are constructed the two witnesses in the same manner, namely the first witness from IX. and X.. and the second from XI. and XII.

WITNESSES.

I.	II.
O	O O
O O	O
O	O O
O	O

and from the two witnesses,

THE JUDGE.

O
O
O O
O

To recapitulate we will seek the answer to the following question :—

Shall I succeed in my undertaking ?

O O	.	. / .	. / .	. /	. /	. / .	.
O O	. / .	/ .	. / .	. / .	. / .		
O	. / .	. /	. /	. / .	. /		
O O	.	/	/	. /	. /	.	
O	/ .	/ .	/ .	. / .	. / .	.	
O O	. /	. / .	/	. /			
O	/ . . /	. / .	.				
O O	,	/ .	/ .	.	.		

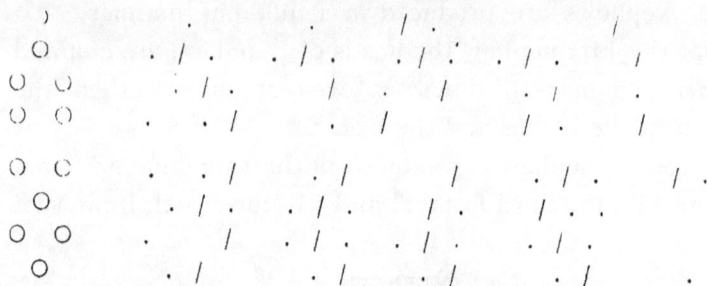

We have, therefore, the

MOTHERS.

I.	II.	III.	IV.
O O	O	O	O O
O O	O O	O	O
O	O	O O	O O
O O	O O	O O	O

DAUGHTERS.

V.	VI.	VII.	VIII.
O O	O O	O	O O
O	O O	O	O O
O	O	O O	O O
O O	O	O O	O

NEPHEWS.

IX.	X.	XI.	XII.
O	O	O O	O
O O	O O	O	O
O O	O O	O O	O O
O O	O	O	O

WITNESSES.

XIII.		XIV.
O O		O
O O		O O
O O		O O
O		O O

JUDGE.

XV.

O

O   O

O   O

O

In this case the Judge is *Carcer*, meaning prison, capture, or attainment of the object desired. It has been made of *Tristitia* and *Laetitia*, meaning sorrow followed by Joy, and the answer is therefore :—

The beginning is painful, but you will succeed in the end.

If analyzed still more we find :

I. is *Albus*, signifying that the undertaking must be begun with wisdom.

II. is *Amissio*, meaning that a sacrifice must be made.

III. is *Fortuna major*, showing that the undertaking, if successful, is well worth the trouble involved.

IV. is an important figure ; it always refers to the termination. Being in this case *Acquisitio*, it means that the object will be obtained, and thus it confirms the decisions of the judge.

Sometimes, if the answer is not satisfactory, a *Supreme Judge* may be constructed out of I. and XV.

The above is a short and easy method, and sufficient to answer simple questions. If, however, any more detailed information is desired, a house is to be constructed according to astrological rules.

ASTROLOGICAL GEOMANCY.

An astrological figure consists of twelve houses, and is constructed in the following manner :—

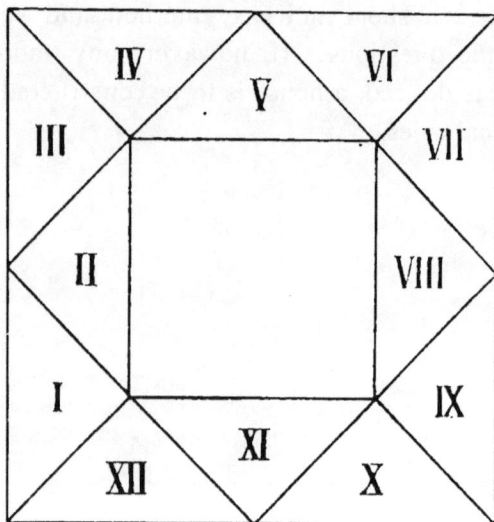

Each of these houses is under the predominating influence of some certain planet or planets, whose action is modified by their position in regard to the rest.

House I. is the house of Life, and therefore represented by ⊙.

II. is dedicated to ☿.

III. The influence of ☾ is mixed with that of ☿ and ♄.

IV. is especially under the rule of ♄.

V. Here the influence of ♀ is predominant, mixed with that of ♄ and ♃.

VI. In this house ♂ is the principle ruler.

VII. is dedicated to ♀.

VIII. to ♄.

IX. to ☾.

X. to the ☉.

XI. Here the influence of ♃ is predominant.

XII. is under the pernicious influence of ♄ in its evil aspect.

House I. deals especially with the person of the questioner, also with matters regarding life, health, appearance, beauty, colour, riches, fortune, success.

House II. with profit and loss, mercantile matters.

House III. Relatives, letters, little voyages.

House IV. Parents, property, treasures, agriculture, mines.

House V. Women, children, luxury, eating, drinking, pleasures, servants, legacies.

House VI. Diseases, servants, misfortunes, domestic animals.

House VII. Women, marriage, whores, thieves, robbers, dishonours.

House VIII. Death, legacies, trouble, suffering, poverty.

House IX. Religious matters, long voyages, dreams.

House X. Fortune, honours, kings, glory, fame, victory.

House XI. Protection, riches, presents, friends, joy, hope, confidence.

House XII. Loss, imprisonment, secret enemies, vagabonds, prostitutes, beggars, misfortune.

The usual way of proceeding is to insert the above-described 15 symbols in the order in which they have been received into the houses which their numbers indicate; namely, the 1st mother in the 1st house, the 2nd in the 2nd, etc., as

may be seen in the figure below: The two witnesses and
the judge are to be put into the middle field :—

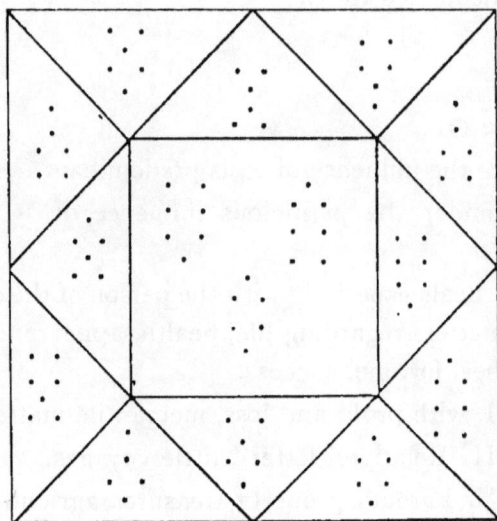

Cornelius Agrippa, however, recommends another method
as being superior to this ; namely, to insert the symbols into
the houses in the following order :—

Mother	I.	into house	I.
,,	II.	,, ,,	X.
,,	III.	,, ,,	VII.
,,	IV.	,, ,,	IV.
Daughter	I.	,, ,,	II.
,,	II.	,, ,,	XI.
,,	III.	,, ,,	VIII.
,,	IV.	,, ,,	V.
Nephew	I.	,, ,,	IX.
,,	II.	,, ,,	VI.
,,	III.	,, ,,	III.
,,	IV.	,, ,,	XII.

The above figure will then present the following aspect:—

Mothers:	1.	2.	3.	4.
	O	O	O	O O
	O	O O	O	O
	O	O	O O	O O
	O O	O O	O O	O

Daughters:	1.	2.	3.	4.
	O O	O O	O	O O
	O	O O	O	O O
	O	O	O O	O O
	O O	O	O O	O

Nephews:	1.	2.	3.	4.
	O	O	O O	O
	O O	O O	O	O
	O O	O O	O O	O O
	O O	O	O	O

These symbols are inserted into the houses as described above.

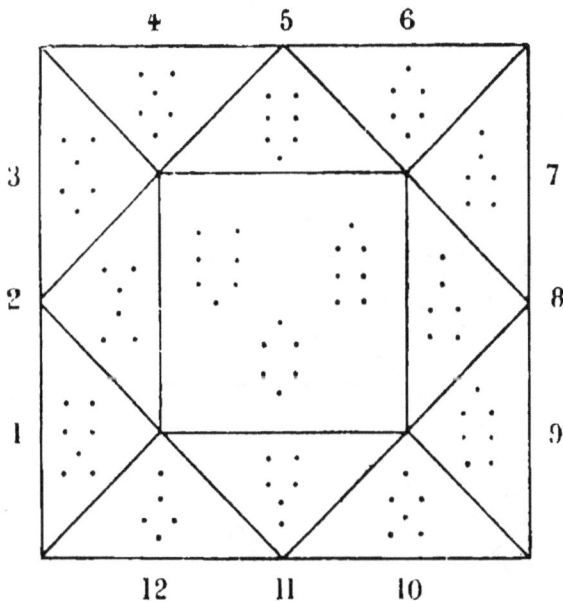

The figure is then to be amplified by adding the signs of the zodiac. The zodiacal sign of the first Mother is put next to House I. The others then follow in their regular order.

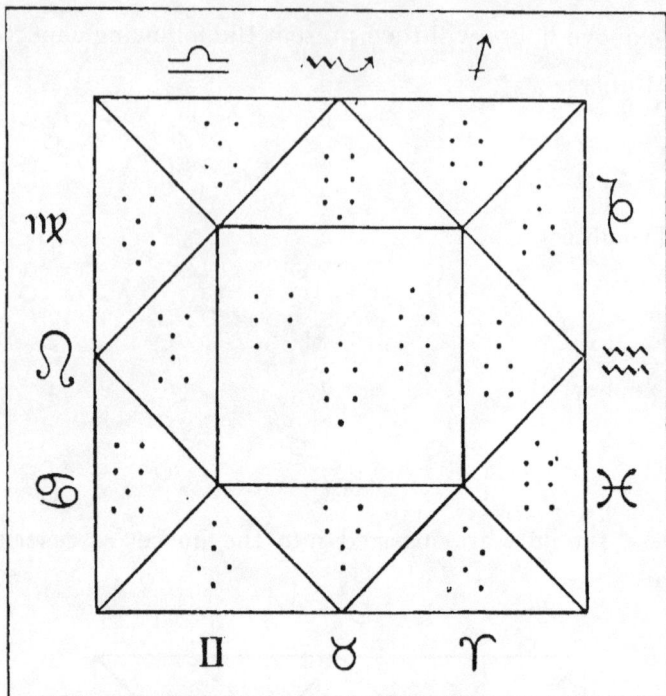

The figure may then be completed by inserting the planetary signs corresponding to each symbol.

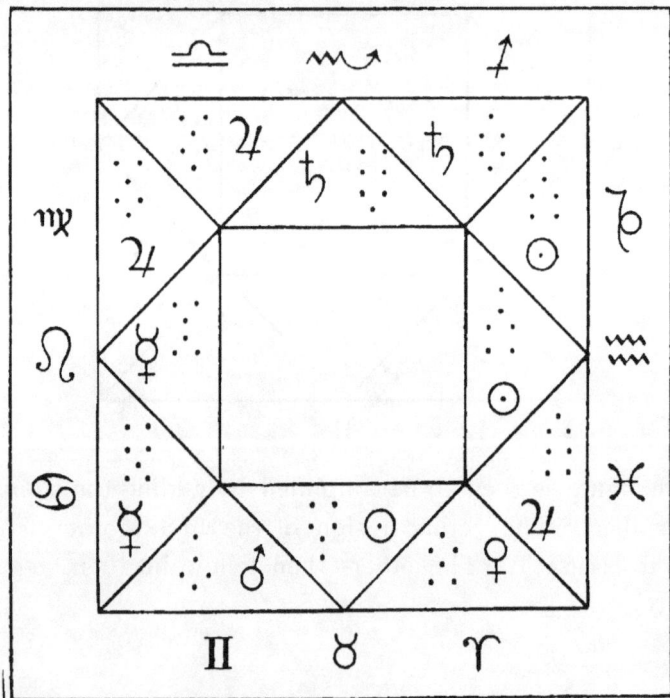

To find the Judge or Indicator according to this method, all the points of the whole scheme received by punctuation are counted together and divided by 12. The remaining points are divided off one by one into the houses, beginning by I. and the house which receives the last point, is the one whose symbol is to be the Judge.

In our scheme on page 37, the number of points amounts to 169. Divided by twelve there remains only one point. The symbol of the first house is therefore in this case the Judge, and his decision becomes intelligible by taking his position to the other houses and planets into consideration. If the aspects of the other planets are doubtful, the opinion of the judge alone is decisive.

SIGNIFICATIONS OF THE GEOMANTIC SYMBOLS ACCORDING TO THEIR POSITIONS.

To practise successfully the art of Geomancy, it is necessary to keep in mind the significations of the planetary signs, the position which they occupy in regard to the signs of the zodiac, the houses in which they reside, and their constellations among each other. To become an experienced geomantist is a difficult undertaking, unless one possesses the necessary talents for it. Geomancy is an art, and, like all other arts, requires to be attained by practice. To facilitate this practice the meaning of the geomantic symbols, according to their position, is now given below:—

FORTUNA MAJOR.

$$\begin{matrix} O & O \\ O & O \\ & O \\ & O \end{matrix}$$

In House I. signifies a noble character, long and happy life, a person of middle size, moral and benevolent.

In II. Riches and gain, being fortunate in recovering a thing that was lost, capture of a thief, etc.

In III. Noble and agreeable relatives, fortunate voyages, friends.

In IV. A noble and respected father, inheritance, success, recovery, success in mines.

In V. Joy from children, honours, fame, birth of a son.

In VI. Health, recovery; a good physician; faithful servants.

In VII. A rich, honest and amiable wife; happy marriage, agreeable love affairs, gaining of lawsuits, and also powerful opponents.

In VIII. The person about whose death one inquires, still lives. It also signifies a painless and natural death, honours after death, legacies and a great dowry belonging to one's wife.

In IX. Extensive but fortunate voyages; safe return; a man faithful in religion; important dreams, visions, intuition; spiritual knowledge.

In X. Great honours; public and honourable positions; upright judges; the lawsuit will be quickly decided; fortunate kings; victory; long life of a noble mother.

In XI. True and useful friends; a rich and benevolent nobleman; success at the king's court; happiness.

In XII. If the question is about enemies, it means that they are powerful and influential. If the question is, whether one will be victorious over one's enemies, it means a lucky escape. It also signifies faithful servants, escape from prison, and that an impending danger will be avoided or not become very serious.

FORTUNA MINOR.

```
      O
      O
   O  O
   O  O
```

In I. Long life, but some troubles or ailments. A man or woman of small stature.

II. Money, but squandering; lavish expenses; the thief remains hidden; stolen articles are not recovered at all, or with great trouble.

III. Trouble and annoyances from relatives; threatening danger on a voyage, but escape therefrom. Reliable but reserved and uncommunicative persons.

IV. Loss to that which has been inherited from the father and other legacies; difficulty in obtaining lost or hidden things. No success in mines.

V. Only few children; a girl is to be born; honourable positions with but little remuneration. Small honours, little fame.

VI. Sanguinic or choleric diseases; the patient is in danger, but will escape; honest but lazy and useless servants.

VII. Marriage with a woman of good family, but some trouble connected with it. Inconstant loves. Procrastination. Tedious law-suits but final success.

VIII. Death in a foreign country; legacies obtained with trouble and vexations; the dowry of the wife will soon be used up, or obtained with difficulty.

IX. Trouble on a voyage; theological occupations; imperfect knowledge.

X. Powerful kings and nobles; gaining possession by force; great honours and positions but instability of fortune. Tedious law suits. Illness not serious.

XI. Many friends, but poor and useless ones. Favours from high personages. Inconstancy of fortune.

XII. Sly and smart enemies; to a prisoner a long captivity but final freedom. Useless servants. Frequent changes.

VIA.

O
O
O
O

In I. Long and happy life. A stranger of tall figure, thin, liberal, agreeable, but not much inclined to labour.

II. Increase of fortune; recovery of lost or stolen property, but escape of the thief.

III. Many brothers and relatives. Many fortunate voyages, sociability.

IV. An honest father; increase of fortune inherited from the father; good harvest; gain. Mines.

V. Numerous male children, a son will be born; honorable positions in foreign countries.

VI. Protection against diseases; the patient will quickly recover; useful servants and animals.

VII. A beautiful and agreeable wife, lasting happiness in marriage; favourable progress of lawsuits, profitable settlements.

VIII. Death in consequence of phlegmatic diseases; great legacies; the person believed to be dead is still living.

IX. Long voyages on the water; profit; clerical positions; profit from church affairs; simple but firm faith: significant dreams; philosophical and grammatical knowledge.

X. Fortunate, kings and noblemen in peace with their neighbours; friendships; public honours; official positions, profitable money affairs; lawsuits will be quickly decided; a respected mother.

XI. Many useful friends; confidence of superiors; business connected with travelling.

XII. Many enemies, doing, however, little harm; useful servants; escape of prisoner, protection in misfortune.

POPULUS.

```
O  O
O  O
O  O
O  O
```

In I. Life of average duration; diseases and changes of fortune. A person of medium stature; thick.

II. Moderate fortune, obtained with much trouble. The stolen property will not be recovered, nor will that which has been lost be completely restored. The thief has not escaped, but is hidden.

III. An average number of relatives; little profit, a vacillating mind; loss by being cheated.

IV. A sickly father; no inheritance of real estate, but profit in things connected with water. Trouble about inheritance. No success in mining.

V. Ordinary, neither very profitable nor very respected positions; calumny, slander, gossip; wife bears no children; miscarriages.

VI. Cold diseases, especially of the lower extremities; a careless physician; danger of death; difficult recovery; dishonest servants.

VI. A beautiful and pleasant wife; but one who is not very faithful. Hypocrisy; pretended loves; impotent enemies.

VIII. Quick death, perhaps by water. No inheritance. Legacies lost by lawsuits. Wife has very little dowry.

IX. Deceptive dreams; a vulgar and coarse person; in clerical matters low positions; indifference towards religion; not very conscientious.

X. Kings and noblemen who lose their positions; loss; offices connected with water; tedious lawsuits; a sickly mother.

XI. Few friends, but many flatterers. No favours to be expected from superiors; weak and ignoble enemies; the prisoner will not escape; danger from water.

ACQUISITIO.

```
  O   O
    O
  O   O
    O
```

I. Long life; happy old age. A man of medium size with

a big head ; marked features ; spends much for himself, but gives little away.

II. Great riches. Lost or stolen goods will be restored.

III. Many relatives with ample means. Many fortunate and profitable voyages. Fidelity and sincerity.

IV. A considerable inheritance from the parents ; great possessions, large harvests ; a hidden treasure ; mines can be found ; a rich, but avaricious father.

V. Numerous children of either sex ; but more boys than girls ; a son will be born ; profitable offices.

VI. Many long and serious diseases ; danger of death ; but an experienced physician. Many servants. Profit.

VII. A rich wife, a widow or advanced in years. Long and tedious lawsuits, a love affair or a concubine.

VIII. The person inquired after is dead. Quick death after a disease of only a few days' duration. Profitable legacies. Rich dowry.

IX. Long and profitable voyages. The absent person will soon return. Profit from theologians and teachers. The person inquired after has considerable knowledge.

X. To kings augmentation of possession. In law a judge who is favourably inclined, but who expects presents. Profitable positions and business. A rich and happy mother.

XI. Many useful and profitable friendships. Favours from high personages.

XII. Many and powerful enemies. Recovery of lost animals. The prisoner will not escape.

LAETITIA.

O

O　O

O　O

O　O

I. Long, fortunate and joyful life. A person of tall stature, fine figure and features.

II. Riches, but also great expenses. Stolen things will be restored; but the thief escapes.

III. Agreeable but short-lived relatives; good voyages; fidelity and sincerity.

IV. Considerable parental fortune; possessions; a noble father. A rich mine may be found.

V. Obedient and good-natured children. A daughter will be born. A good reputation.

VI. The patient recovers. Useful servants.

VII. A young and beautiful wife; gaining of lawsuit; fortunate in love affairs.

VIII. Legacies. The person inquired after is still living.

IX. Few voyages. A man of a religious character, not very learned, but intuitive.

X. Kings and nobles of peaceful character. Honourable positions in the church or in law. If the mother is a widow, she will marry again.

XI. Many friends among the high. Protection.

XII. Victory over enemies; useful servants; freedom for the prisoner, protection against evils.

PUELLA.

O

O O

O

O

I. Rather short life. A man of middle size and feeble constitution, of a feminine character, full of sensual desires, and who often gets into trouble on account of his love for the other sex.

II. No increase of riches nor greater poverty. Lost or stolen things will not be restored. The thief has left the city.

III. More sisters than brothers. Agreeable voyages. Pleasant social surroundings.

IV. The inherited fortune is small. The harvest will be good.

V. The expected child is a girl. Favours received through the influence of ladies.

VI. The patient is very feeble, but will speedily recover. The physician is ignorant and inexperienced, but the vulgar people have great respect for him. Useful servants.

VII. A beautiful and agreeable wife, living in peace with her husband, but being of an amorous nature and having many admirers. No serious law-suits of any kind.

VIII. The person believed to be dead still lives. The dowry is small, but the man is satisfied with it.

IX. Short voyages. A religious-minded man without great talents, except for music and singing.

X. Powerful and peaceful kings and noblemen loving sport. Honest judges. Positions with ladies of rank.

XI. Many friends among men and women.

XII. Only few enemies; but trouble with women. The prisoner will obtain his freedom through the influence of his friends.

AMISSIO.

O
O O
O
O O

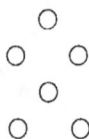

I. The patient will not recover. A short life. A man of irregular form, spiteful and disagreeable, having some blemish, such as a squint or limping, etc.

II. Loss or squandering of money. Poverty. That which is lost or stolen will not be restored; the thief will escape. No luck in mining.

III. Few relatives, or death of the latter. No important voyages. A great deal of cheating.

IV. The inheritance from the father is rapidly lost. The father is poor, and dies suddenly.

V. Death of children. Miscarriages. Neither honours nor fame, but a great deal of slander.

VI. The patient will recover. Useless servants. Misfortune with domestic animals.

VII. An adulterous and quarrelsome wife, who, however, will die soon. Loss of lawsuits.

VIII. Death of an acquaintance. No legacies, or loss of the latter.

IX. No voyages, or if there are any they will be unfortunate. A person of vacillating mind, changing his belief frequently. A person ignorant in every respect.

X. Unfortunate kings or nobles, ending in exile or losing their positions. Ignorant judges, or such as can be bribed. Positions that will cause loss and harm the reputation. Death of the mother.

XI. Only few friends. Friendship easily lost. Favours, if any, will bring no profit.

XII. The enemies will be annihilated. The prisoner will be long in captivity, but is otherwise safe.

CONJUNCTIO.

$$
\begin{array}{cc}
\bigcirc & \bigcirc \\
& \bigcirc \\
& \bigcirc \\
\bigcirc & \bigcirc
\end{array}
$$

I. Long life. A man of medium size; face long, agreeable and having many friends.

II. Neither riches nor poverty. The thief will be caught. Lost or stolen property will be returned. Success in mining.

III. Few relatives. Various voyages with changeful success. Reliability of character.

IV. Average fortune from parents; a good and intelligent father.

V. Intelligent children. The expected child is a son. Self-acquired honours, great fame, good reputation.

VI. Long and tedious diseases. An experienced physician; faithful servants.

VII. A well-educated and intelligent wife. Difficult lawsuits with sly opponents.

VIII. The person inquired after is dead. An advantage to be derived from the death of a relative or friend.

IX. Few but long voyages. Knowledge of secrets in religion. An active mind.

X. Good and liberal-minded kings ; upright judges ; positions connected with instruction in natural sciences. A good and intelligent mother.

XI. Many friends and especially great favours from high personages.

XII. The enemies are prudent and sly. The prisoner remains in his prison. Escape from various dangers.

ALBUS.

O   O
O   O
  O
O   O

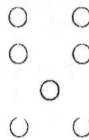

I. A person troubled with continual ailments or serious diseases. A person of small stature, a great talker, gay and amusing.

II. Gain from things that serve for amusement. Discovery of lost or stolen things. No success in mining.

III. Only few relatives ; few but difficult voyages. A great deal of cheating.

IV. Little or no inheritance from the parents.

V. No children, or if there are any, they die. Miscarriage or birth of monstrosity. Slander and gossip. No honours to be expected.

VI. Tedious diseases. Useless and dishonest servants. The patient mistrusts his physician.

VII. A beautiful and beloved wife, but who will bear no children. Few but long-lasting lawsuits.

VIII. The person inquired after will die. The dowry of the wife is small, and will be the cause of a lawsuit.

IX. Voyages bringing but little profit. Obstacles. The absent person will not return. A superstitious man, adhering to false sciences.

X. No favours to be expected from kings or judges. Unprofitable positions or business. The mother is unchaste, or suspected of adultery.

XI. False friends. Hypocrisy, inconstant fortune.

XII. The enemies are impotent. Perversities of various kinds. The prisoner will not escape.

PUER.

O

O

O O

O

I. Life not very long, but full of trouble. A man of strong constitution. An excellent soldier.

II. Money not inherited but acquired. Escape of the thief. No success in mining.

III. Superiority. Dangerous voyages. Good reputation.

IV. Doubtful legacies and possessions. Irregularly acquired riches.

V. Good children, who will become prosperous. The expected child is a son. Military honours. Considerable fame.

VI. Serious diseases. Wounds ; injuries ; but easy recovery. A physician well versed in surgery. Useful servants.

VII. An honest and courageous wife, an order-loving housewife. Difficult lawsuits.

VIII. The person inquired after is alive. Death rapid. No legacies are to be expected.

IX. Dangerous but successful voyages. A person neither

very religious or conscientious. Considerable learning in natural sciences, medicine and arts.

X. Powerful and victorious kings. Changes of fortune. Unmerciful judges. Positions in the army or in professions having to do with iron or fire. Danger to the mother.

XI. Friendship with nobles, especially in the military. Doubtful profit.

XII. Cruel and dangerous enemies. The prisoner escapes. Evasion of dangers.

RUBEUS.

I. Short life, and a bad end. A vicious, cruel and useless person ; a villain ; being especially marked on some part of his body.

II. Poverty ; thieves ; robbers ; counterfeiters; cheats. The thief escapes, and there is no fortune in mining.

III. Hated relatives ; dangerous voyages ; treachery.

IV. Loss of inheritance ; bad harvest; sudden death of the father.

V. Numerous but bad and disobedient children.

VI. Mortal diseases or wounds. The patient dies. The physician makes a mistake. Treacherous servants.

VII. A wife of ill repute, adulterous and quarrelsome. The enemies are treacherous, and will by some trick get the best of the querant.

VIII. Death forcible, in consequence of judicial decision, execution, hanging, etc. The person inquired after is dead. The wife has no dowry.

IX. Dangerous and difficult voyages. Robbery or imprisonment. A person of very little religion ; one who does keep his promises ; unfaithful. False and deceptive sciences.

X. Cruel tyrants, who will die a miserable death. Judges

who must be bribed. Cheats, swindlers, thieves and usurers. The mother dies suddenly, and leaves a bad reputation behind.

XI. Intercourse with bad and disreputable people. Expulsion from good society.

XII. Cruel enemies and traitors. The prisoner will perish. Many obstacles and perversities.

CARCER.

I. A short life. A vicious, ugly, and unclean person, who is an object of hate and contempt.

II. Extreme poverty. The thief will be captured.

III. Dislike among relatives. Evil company. Unfortunate voyages.

IV. No legacies to be expected. The father is a bad man, and will take an evil end.

V. Bad children. The woman is not pregnant. Miscarriage or infanticide. No honours ; but much gossip.

VI. The patient has a long-lasting disease. The physician is ignorant. Bad and useless servants.

VII. The wife is hated by her husband. The lawsuit will be lost.

VIII. Death by a fall or by execution. Suicide. Neither dowry or legacy to be expected.

IX. The absent person will not return, having met with an accident on the way. A person devoid of religious sentiment. A very bad conscience. No culture.

X. Vicious kings and nobles, using their power for the gratification of evil desires ; and who will take a bad ending. Falsifying judges and lawyers. A dishonest, adulterous mother. The person obtains neither honour nor position, but lives by begging, theft or robbery.

XI. No friends nor protectors.

XII. Enemies. The prisoner will not escape. A great deal of misfortune.

TRISTITIA.

```
O   O
O   (
O   (
  O
```

I. The life is not necessarily short, but full of trouble. A good-natured person, but slow in everything; of an eccentric character; melancholy and avaricious.

II. Fortune and riches; but little benefit resulting from its possession, as it is not used, but hidden away. The thief escapes, and the stolen goods are not restored.

III. Few relatives, who will all die before the questioner. Unfortunate voyages.

IV. The expected legacy or possession will not be obtained. A very long-lived and avaricious father.

V. There are no children, or, if any, they will die young. The expected child is a girl. Honours and fame are small.

VI. The patient must die. Faithful but lazy servants.

VII. The wife will die soon. No advantage from lawsuits.

VIII. Death after a long and painful sickness. Legacies. The wife has a dowry.

IX. The absent person is dead, or has met with an accident. Unfortunate voyages. A man of good religion, and possessing considerable knowledge.

X. Severe but just kings and judges. Slow decisions in law. Many obstacles. The mother will have a long life, but various troubles. The positions obtained are important, but not of long duration. Occupation with water or agriculture, or with theological or philosophical matters.

XI. Few friends, and the death of the same.

XII. No enemies. The prisoner will be condemned. Battle with many difficulties.

Caput Draconis.

⌣ ○
○
○
○

I. Long life and fortune.

II. Riches. The thief escapes. Rich mines.

III. Several brothers. Voyages. Relations by marriage.

IV. Rich legacies. The father has a long life.

V. Many children. The expected child is a son, or there may be twins. Honour and fame.

VI. Diseases. An experienced physician. Many servants.

VII. Several marriages. Numerous law-suits.

VIII. Certain death. Legacies. A good dowry.

IX. Many voyages. Safe return. Religiosity. Knowledge.

X. Celebrated kings. Respected judges. A noble mother. Important affairs and remunerative occupations.

XI. Many friends and the favour of all.

XII. Many enemies. Many female acquaintances. The prisoner will not escape, but receive a severe punishment.

Cauda Draconis.

○
○
○
○ ○

This figure signifies in all its houses the exact opposite to that of the preceding one.

Note.

In practising geomancy it is necessary to seek the answer to the proposed question not merely by studying the figure in the house to which that question belongs; but also to take all the figures connected with it into consideration. The principal point in this art is the comparison of the various symbols and of their true signification, an art which is only possible to those who can call to aid the power of their own intuition.

EXAMPLE.

Question : Will the proposed marriage be a success ?

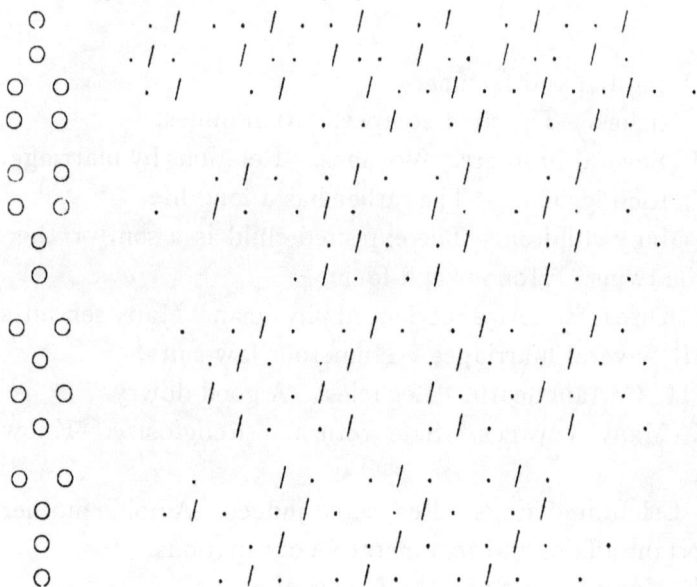

:= 152 points.

MOTHERS.

I.   II.   III.   IV.

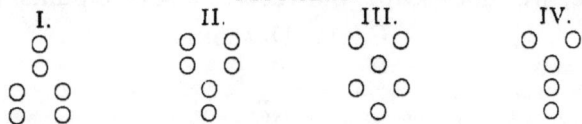

DAUGHTERS.

V.   VI.   VII.   VIII.

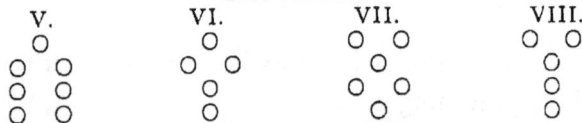

NEPHEWS.

IX.   X.   XI.   XII.

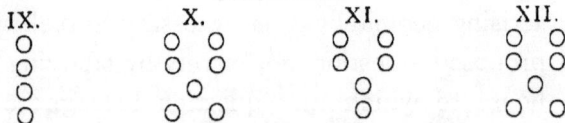

WITNESSES.

XIII.   XIV.

JUDGE.

XV.

○
○
○ ○
○ ○

The answer is : *He will obtain her.*

If we desire to see this statement corroborated, and to know more particulars about the proposed marriage, we may construct an astrological house, and insert the symbols according to the directions given above :

There have been 152 points obtained by punctuation. If we divide this number by 12, there will still remain 8 ; 8 is, therefore, in this case the Judge or indicator. The symbol in house VIII. is *Acquisitio.* The answer is, therefore : *He will obtain her.*

Moreover, Acquisitio in the eight house means that the woman is rich, and a widow ; the sign is ♃, which means that there is a strong power active to produce the desired result. The termination is the fourth house, as indicated by ☋, is very favourable, and predicts that she will inherit a fortune from her father. The ☉ in the house of life (I.) predicts a long and happy life, but not without some trouble ; ☊ in house I. indicates that children will be the result of the marriage. If the whole figure is carefully studied, according to the directions given above, numerous other details may be learned therefrom.

If we stand before a picture representing a landscape, we do not look at merely one house or one tree, to the neglect of the rest ; but we see the whole picture at once, and then we may afterwards examine it in its details. Thus, to the experienced geomantist, such a table of figures represents a picture in which he can at once see the future life of the questioner represented in allegorical figures. As a whole, it has been grasped by his own internal spiritual perception during the process of punctuation. It has been resolved in its details by the science of Geomancy, and it now comes to his external consciousness, as a whole, by the power of his trained intellectual perception.

The soul may know a great deal, while man may know very little ; the former sometimes communicates some of its knowledge to the latter by dreams and visions, when the mind is in that tranquil state between waking and sleeping. The mind is like a mirror, in which not only the images of external things, but also those projected by the soul may be reflected. If the Spirit retires within itself, the mind resembles an unclouded mirror; but as soon as the brain begins to think, the mirror becomes like a pool of water, whose surface is disturbed by falling rain. Man sees then only his own thoughts and not the thoughts of the soul. Those who desire to practice successfully the art of Geomancy should, during the process of punctuation, keep their own thoughts in abeyance, and let the divine soul do its thinking in them.

ASTRONOMICAL GEOMANCY.

There is still another method of practicing Geomancy described by Cornelius Agrippa. It is called the Astronomical Geomancy of Gerhard of Cremona. It is a very simple method, but to be successfully applied it requires a considerable knowledge of the character of the various planets and their constellation. The figure used is similar to those described above.

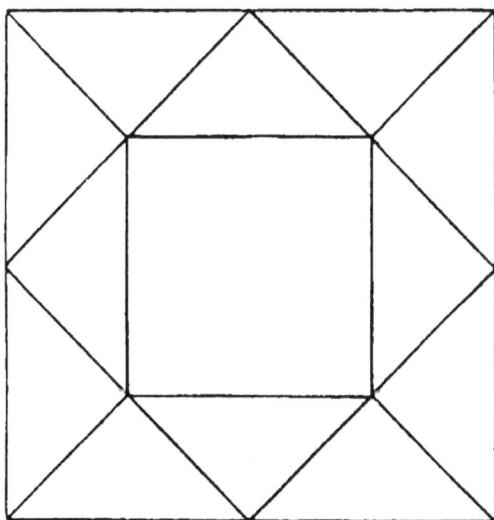

Four lines of points are made in the usual manner, and a geomantic symbol obtained. The zodiacal sign corresponding to it is then put into the first house, and the remaining signs into the other houses, from I. to XII., in their regular order. For instance, if the symbol obtained is *Populus*, its corresponding sign, ♑, is put into the first house, ♒ in the second, ♓ in the third, and so forth to the end. The sign in the first house is called the *Ascendant*.

After this there are again four lines punctuated. The number of points is counted together and divided by twelve. The rest signifies the number of the house into which the first planet, ☉, is to be inserted.

In the same manner the house for the next planet ☾ is to be found, and the same process is repeated for all the planets in the following order :—

$$☉, ☾, ♀, ☿, ♄, ♃, ♂, ☋, ☊.$$

The houses and the planets and signs contained therein are then studied and compared together, as has been indicated above.

The manner of proceeding may be illustrated by the following example.

Question. Will the patient recover ?

```
 O      · / ·   · / ·   · / ·     / ·   · / ·
 O      · / ·    / ·  · / ·  · / ·  · / · ·
 O  O ·  · /    · / ·  · /    · / ·  · / ·
 O      · / ·    / ·  · / ·  · /     · / ·
```

The corresponding zodiacal sign is ♈. It is to be inserted into the first house and the other signs follow in regular order.

We now punctuate again to find the house of the Sun :—

```
I. ·    · / ·      ·   ·   ·           ·
    ·   ·   ·       · / ·   ·   ·       ·  ·
  ·  ·  ·  ·  ·   ·     · / ·   ·   ·
    ·   ·      ·   ·     ·    / ·  · · ·
```

After dividing the number of these points by 12 there remains 9, and ☉ is therefore inserted into the ninth house. Then follows :—

```
II. ·    · / ·  ·   ·   ·           ·
          · / ·     ·   ·   ·   ·
        · / ·        ·   ·    ·     ·
       · · /    ·      ·    ·   ·  ·
```

There remains two, and the ☾ belongs to the second house.

III.

There remain 12, and ♀ is to be put into XII.

In this way the process is continued until the houses for all the planets are found. If two or more planets receive the same number, they are to be put into the same house.

The figure may then represent the following aspect:—

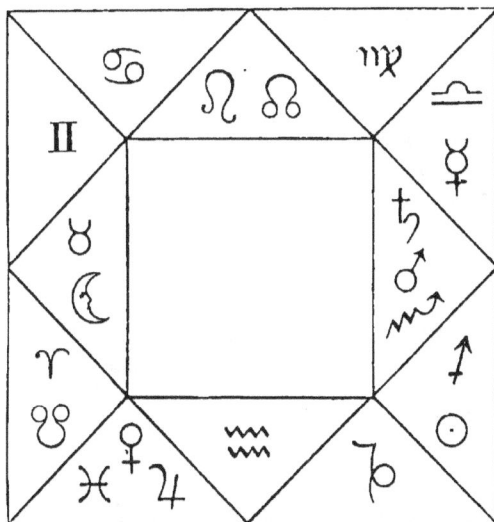

In this case the answer would be, that there is not the least possibility for the recovery of the patient. The presence of ☊ in I. is a very bad sign, indicating an ignorant physician; but the presence of ♄ and ♂ under the influence of ♏ in the house of death is extremely bad, indicating death by iron or fire, and the probability is, therefore, that he will be killed by a surgical operation.

In this way numerous complications may arise from the positions of the various planets in the twelve houses, and those who desire to enter deeper into this subject will find information in regard to the meaning of the different constellations in books on Astrology. Those, however, whose intuition is sufficiently developed to grasp the true

E

meaning of the symbols and signs, will require no further instruction.

CONCLUSION.

If Geomancy was a truth once, it must still be true, and the answer to the question why it is so little practised at present can only be either that man's intuition has been unfolded so much as no longer to require such artificial aids ; or that he has become so much less intuitive as not to be able to grasp truths spiritually, even with such artificial aid.

That the first answer cannot be the true solution of the riddle is self-evident, for there are few persons existing to-day who can correctly predict any future event, unless they base their prophecies upon calculations made from external observation. If the attention of the reader has been called to the existence of a higher power than the perishing intellect, and he has been encouraged to put more faith in the knowledge of God, the object for which these pages were written will have been accomplished.

APPENDIX,

CONTAINING 2,048 ANSWERS TO QUESTIONS.

—

TRANSLATED

FROM THE GERMAN OF THE 16TH CENTURY.

NOTICE.

The following answers are obtained by constructing the I. and II. Witness, and subsequently the Judge according to the directions given on page 38 of the preceding part of this book. The middle figure is the Judge; the left-hand figure is the first, and the right-hand one the second Witness.

QUESTIONS.

1. Will the person inquired about have a long life ?
2. Will he become rich ?
3. Will the proposed undertaking succeed ?
4. How will the undertaking end ?
5. Is the expected child a boy or a girl ?
6. Are the servants honest ?
7. Will the patient recover ?
8. Will the lover succeed ?
9. Will the inheritance be obtained ?
10. Will the lawsuit be gained ?
11. Will the desired position be had ?
12. What will be the kind of death ?
13. Will the expected letters arrive ?
14. Will the voyage be fortunate ?
15. Will good news arrive ?
16. Will the adversary be conquered ?

INDEX.

♃
Acquisitio.

○ ○
○
○ ○
○

☾
Populus.
○ ○
○ ○
○ ○
○ ○

Acquisitio.

♃

1. A long and happy life.
2. Will be rich in old age.
3. It will be successful.
4. The object will be slowly attained.
5. A boy.
6. Do not trust them too much.
7. He dies after a long sickness.
8. Success.
9. Yes, but it will take a long time.
10. Money will gain the day.
11. No success.
12. Dropsy.
13. Yes.
14. Fortunate, but slow.
15. Good news will arrive.
16. Yes.

☾
Populus.
○ ○
○ ○
○ ○
○ ○

○ ○
○
○ ○
○

♃
Acquisitio.
○ ○
○
○ ○
○

Acquisitio.

♃

1. Feeble in youth, afterwards strong.
2. Wealthy in youth, afterwards poor.
3. Do not doubt.
4. The end is good.
5. Son.
6. Reliable, but greedy.
7. He will die soon.
8. After waiting a long time.
9. At last something will be had.
10. The judge is favourably inclined.
11. The aspect is favourable if it refers to lawyers.
12. Death by water or watery diseases.
13. Slowly.
14. Fortunate.
15. Tolerable.
16. Yes.

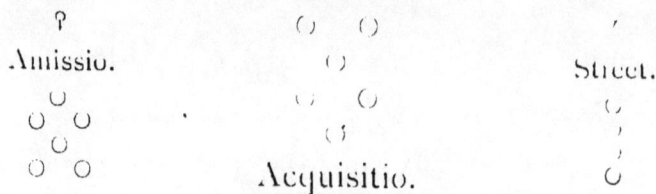

Amissio.   Street.

Acquisitio.

♃

1. He will die early.
2. Inconstant fortune.
3. Many disappointments.
4. Difficult beginning but a good end.
5. Daughter.
6. Worthless.
7. Dies.
8. He may obtain her by strategy.
9. The legacy is bad.
10. Loses.
11. Inconstant fortune.
12. Fever.
13. Yes.
14. In no way favourable.
15. Trifles.
16. The adversary is stronger than you.

Street.   Amissio.

Acquisitio.

♃

1. Average.
2. A little in the beginning, but nothing in the end.
3. Doubtful.
4. Tolerable.
5. Daughter.
6. They are thieves.
7. Dies.
8. She is inconstant.
9. The inheritance is not good.
10. Success through women.
11. Fortunate with ladies of high position.
12. The fever will kill him.
13. No.
14. A good voyage on land in female company.
15. Only tolerable news.
16. Victory.

♃
Laetitia.

♂
Puer.

Acquisitio.

♃

1. Feeble in youth, strong in old age.
2. Final success.
3. Success.
4. The beginning bad, the end good.
5. Son.
6. They are faithful.
7. Recovers.
8. Embrace.

9. Good inheritance gained by law.
10. Loses.
11. Fortunate with clergymen.
12. Febrile diseases.
13. No.
14. Beware of robbers.
15. Bad news.
16. The adversary gains.

♂
Puer.

♃
Laetitia.

Acquisitio.

♃

1. Strong in youth, feeble in old age.
2. That which is inherited will nearly all be lost
3. Fails.
4. A good beginning and a bad end.
5. Son.
6. Reliable.
7. Recovers.

8. He will obtain his wish.
9. A good inheritance.
10. Gains.
11. Success with clergymen.
12. Congestion or apoplexy.
13. They will arrive.
14. No obstacles.
15. News about advancement in office.
16. The adversary loses.

♄
Tristitia.

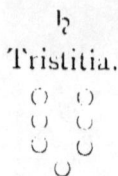

Acquisitio.

♃

♂
Rubeus.

1. Short and painful life.
2. Be not deluded.
3. Cease to think of it.
4. Bad end.
5. Son.
6. They steal secretly and openly.
7. He will have to go.
8. Vain hope.
9. Obtains.
10. Loses.
11. Unlucky.
12. Cold steel.
13. Waits in vain.
14. Wounded during the voyage.
15. Bad news.
16. The adversary gains by cheating.

♂
Rubeus.

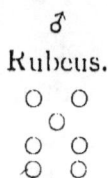

Acquisitio.

♃

♄
Tristitia.

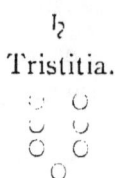

1. Painful and feeble.
2. What you gain will be taken away.
3. The undertaking will be abandoned.
4. Bad ending.
5. Son.
6. Thieving and unreliable.
7. Dies.
8. Loss on account of neglectfulness.
9. The legacy is worthless.
10. Gains, but receives no advantage from it.
11. No luck except in war.
12. Beware of fire and lead.
13. The messenger does not arrive.
14. The voyage will cost his life.
15. Worthless.
16. Victory by strategy.

☊
Caput Draconis.

O O
O
O O
O

☿
Albus.

Acquisitio.

♃

1. Long life.
2. Fortune acquired by writing.
3. Fortunate.
4. Good end.
5. Daughter.
6. Honest and industrious.
7. Recovers after a long illness.
8. Will ultimately succeed.

9. A fat legacy.
10. Settlement.
11. Fortunate with mercurial people.
12. Dies quietly in bed.
13. Soon.
14. Good but slow.
15. Agreeable.
16. Adversary is feeble and desires to settle.

☿
Albus.

O O
O
O O
O

☊
Caput Draconis.

Acquisitio.

♃

1. Happy old age.
2. A fortune by marriage and employés.
3. Keep your undertaking secret.
4. The end is as it is desired.
5. Daughter.
6. Faithful and discreet.
7. Danger.
8, He will obtain her.

9. Slow success.
10. The adversary has the advantage.
11. Fortunate with clergymen and ladies.
12. Natural death.
13. Soon.
14. Fortunate.
15. Good.
16. You will wish to settle,

℧
Canda Draconis.

○ ○
○

♀
Puella.

○
○
○
○ ○

○ ○
○
·○

○
○ ○
○
○

Acquisitio.

♃

1. Not very long. Feeble old age.
2. Means by marriage.
3. Fortunate in love affairs.
4. The end very doubtful.
5. Son.
6. Unchaste and thieving.
7. Dies.
8. Obtains her.
9. Obtains the inheritance.

10. Gains through the influence of women.
11. Fortunate with ladies.
12. Dies of venereal disease.
13. The letters will arrive.
14. Happy voyage, but great expenses.
15. News about love affairs.
16. You will be victorious over your malicious enemy.

♀
Puella.

○ ○
○
○ ○
○

℧
Canda Draconis.

○
○ ○
○
○

○ ○
○
○

○
○
○
○ ○

Acquisitio.

♃

1. Feeble youth, happy old age
2. Your wife will cause your ruin.
3. If the object is robbery, it will succeed.
4. The end is better than the beginning.
5. Son.
6. Thievish and unchaste.
7. Dies.
8. There are enemies in the way.

9. The legacy will go into the hands of strangers.
10. You will lose because ♀ is against you.
11. No luck.
12. Beware of being wounded.
13. No letters.
14. The voyage is very dangerous.
15. News about war.
16. Your enemy is stronger than you.

Fortuna major.

Acquisitio.

Conjunctio.

♃

1. Long life.
2. Rich by trade.
3. Doubtful.
4. Tolerable.
5. Daughter.
6. Faithful.
7. Dies.
8. Certain success.
9. Obtains the legacy.
10. Settlement.
11. Fortunate by means of the pen.
12. Dies in his bed.
13. Soon.
14. Quick and fortunate.
15. Advantageous.
16. You will conquer.

Conjunctio.

Acquisitio.

Fortuna major.

♃

1. Happy and very old age.
2. Great riches.
3. Good success.
4. Good.
5. Daughter.
6. Useful.
7. Regains his health.
8. Success by negotiation.
9. Obtains an excellent inheritance.
10. Gains.
11. Fortunate in high places.
12. Natural death.
13. Not as soon as expected.
14. Fortunate but slow.
15. Happy news.
16. Meet him without fear,

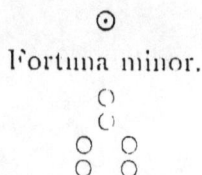

Fortuna minor.

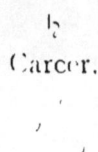

Carcer.

Acquisitio.

♃

1. Average.	9. The legacy is poor.
2. Little gained by a great deal of trouble.	10. The adversary gains.
3. You will cease to think of it.	11. Fortunate in the country.
4. Success slow.	12. Dies of a cold.
5. Daughter.	13. No letters.
6. Lazy and vain.	14. Slow and a great deal of of annoyance.
7. Dies.	15. The news are worthless.
8. Yes, if she is a widow.	16. The enemy will be victorious.

1. Average.
2. Little gained by a great deal of trouble.
3. You will cease to think of it.
4. Success slow.
5. Daughter.
6. Lazy and vain.
7. Dies.
8. Yes, if she is a widow.
9. The legacy is poor.
10. The adversary gains.
11. Fortunate in the country.
12. Dies of a cold.
13. No letters.
14. Slow and a great deal of of annoyance.
15. The news are worthless.
16. The enemy will be victorious.

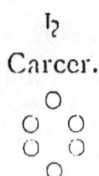

Carcer.

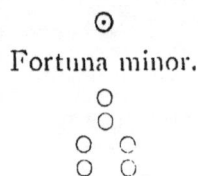

Fortuna minor.

Acquisitio.

♃

1. Average.
2. Fortune through ladies.
3. Final success.
4. Tolerable.
5. Son.
6. Vain and idle.
7. Recovery.
8. Success.
9. The adversary obtains the legacy.
10. Gain.
11. Some success.
12. Dies of a fever.
13. No letters.
14. Tolerable.
15. Letters from high personages.
16. You will not gain.

Amissio.

1. Short and feeble life.
2. Small fortune.
3. Disappointment.
4. Unsuccessful.
5. Boy.
6. Quick, but unchaste.
7. Dies.
8. Loses her on account of a long voyage.
9. Another person obtains the legacy.
10. Neither one of the opposing parties will have an advantage.
11. Unfortunate.
12. Danger of drowning.
13. Soon.
14. He will go to see his sweetheart.
15. News about voyages.
16. The enemy will do no harm.

Amissio.

1. Feeble and short life.
2. Fortune easily lost.
3. Untimely love is injurious.
4. A bad end.
5. Son.
6. Careless and unfaithful.
7. Recovery.
8. He will obtain his wish.
9. The legacy will be very meagre.
10. The opponent has the advantage.
11. Success by being genteel.
12. Danger of death by venereal disease.
13. Delay.
14. Interference from water.
15. Contents are love affairs.
16. You will conquer all enemies.

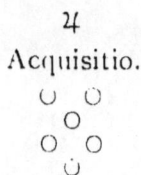

Acquisitio.

Amissio.

Via.

♀

1. Average.
2. Poor in youth and old age. Rich in middle years.
3. The chances are against it.
4. Tolerable.
5. Daughter.
6. Good but inconstant.
7. He regains his health.
8. Your inconstancy will cause you to lose her.
9. Not by you.
10. You will lose.
11. Fortunate on a voyage.
12. Dies in a foreign country.
13. Receives letters.
14. Rapid voyage on land and water.
15. Unimportant.
16. Powerful enemies.

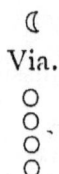

Via.

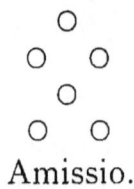

Amissio.

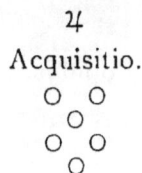

Acquisitio.

♀

1. Healthy youth, feeble old age.
2. The older he becomes, the poorer will he be.
3. The prospect is good.
4. Good.
5. Boy.
6. Yes.
7. A long disease.
8. You will get her.
9. Legacy obtained through females.
10. Victory.
11. You will be welcome.
12. Peaceful and natural.
13. Long delay.
14. Delayed.
15. Joyful contents.
16. You are superior to him.

♃
Lactitia.

☉

○   ○
○
○   ○

Amissio.
♀

♉
Albus.

○   ○
○   ○
○
○   ○

1. Long, joyful life.
2. The older he becomes the richer he will be.
3. Very good prospect.
4. The end will show your ability.
5. Girl.
6. Honest and useful.
7. Dies.
8. Will get her through writing.
9. Another one will get the best part.
10. Amicable settlement.
11. Your pen will be your recommendation.
12. Natural death.
13. They have already arrived.
14. Fortunate.
15. Good.
16. You should try to obtain a settlement.

☿
Albus.

☽

○   ○
○
○   ○

Amissio.
♀

☿
Lactitia.

○
○   ○
○   ○
○   ○

1. Long life and health.
2. Rich by service.
3. The wish will be fulfilled.
4. The beginning is better than the end.
5. Son.
6. Reliable and good.
7. Health.
8. Will get her.
9. You will get the best part.
10. Gains.
11. Honourable positions in the church.
12. Peaceful and easy death.
13. Certainly.
14. Very good.
15. Agreeable.
16. The enemy will do no harm.

Page 84

GEOMANCY.

Tristitia.

Amissio.

Puella.

1. Average duration. Troublesome.
2. Means by way of marriage.
3. Yes, if it is a love affair.
4. Tolerable.
5. Son.
6. Not very faithful.
7. Will gain strength.
8. Most certainly.
9. You may hope for a little.
10. Gains through ladies of rank.
11. All favours by female influence.
12. Melancholy.
13. Letters from ladies.
14. Tolerable.
15. The contents are love matters.
16. Look out for treachery.

Puella.

Amissio.

Tristitia.

1. Long but troubled life.
2. A little by agriculture and hard labour.
3. Useless.
4. Worthless.
5. Girl.
6. Idle and dishonest.
7. Recovery.
8. Lost on account of carelessness.
9. A little, but not much.
10. The adversary gains.
11. Unlucky in the country.
12. Gloom and discouragement.
13. No letters.
14. Slow.
15. Bad news.
16. The enemy gains the day.

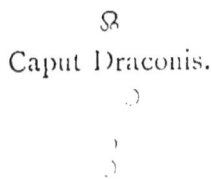

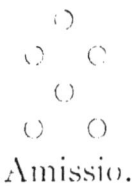

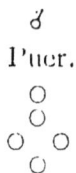

Amissio.

♀

1. Sickly in youth; afterwards stronger.
2. Nothing to be expected except by force.
3. Unsuccessful.
4. Bloody.
5. Son.
6. Average.
7. Will gain strength.
8. The marriage will take place.
9. There is nothing to be inherited.
10. The adversary gains.
11. Fortune in war.
12. Beware of fire.
13. Soon.
14. Dangerous.
15. News about war.
16. Your enemy is too sly for you.

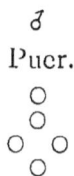

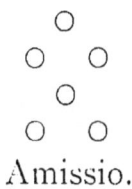

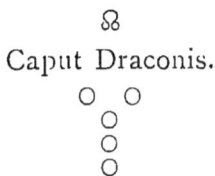

Amissio.

♀

1. Strong in youth, feeble afterwards.
2. A little by marriage.
3. This time you will have success.
4. Fortunate end.
5. Girl.
6. Good.
7. Recovers his strength slowly.
8. The marriage will not take place.
9. A small legacy.
10. Fortunate ending.
11. Fortunate at the court.
12. Natural death.
13. Slowly.
14. Fortunate.
15. Very agreeable.
16. You have the advantage.

℧

Cauda Draconis.

( )

○ ⸝

☌

Rubeus.

Amissio.

♀

1. Long life. Feeble old age.
2. Considerable means.
3. Good success.
4. Fortunate end.
5. Daughter.
6. Average.
7. Dies.
8. It will come to nothing.
9. A small amount.
10. It will not be to your advantage.
11. A good position at the Court.
12. Taking cold.
13. Soon.
14. Fortnnate.
15. Good.
16. You are superior to the other.

♂

Rubeus.

Amissio.

♀

℧

Cauda Draconis.

1. Feeble youth, heathy old age.
2. Obtains some means in old age.
3. It will be abandoned.
4. Unfortunate.
5. Son.
6. Worthless.
7. Dies.
8. The marriage will take place.
9. Another person will get it.
10. The adversary has the advantage.
11. Go to the country or to the war.
12. A fever will make an end to his life.
13. Very slowly.
14. There is dangers from robbers and lewd women.
15. Disagreeable news.
16. Your enemy is too powerful for you.

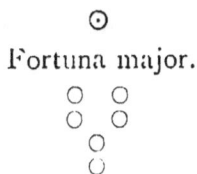

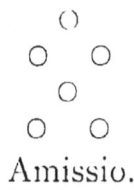

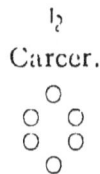

⊙

Fortuna major.

○ ○
○ ○
○
○

( )

○ ○
○
○ ○

Amissio.

♄

Carcer.

○
○ ○
○ ○
○

♀

1. A long life.
2. Means by agriculture and mines.
3. That which seems difficult now will become clearer.
4. Average and slow success.
5. Girl.
6. Good and useful.
7. Dies.
8. You ought to dismiss all thoughts of marrying.
9. The legacy is destined for you.
10. The adversary will receive the judgment.
11. Your fortune is to be found in the country.
12. A natural death.
13. No letters.
14. Fortunate but slow.
15. Letters on agricultural matters.
16. You had better remain away.

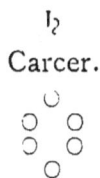

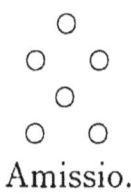

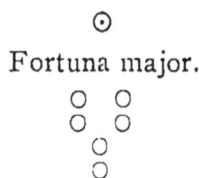

♄

Carcer.

○
○ ○
○ ○
○

○
○ ○
○
○ ○

Amissio.

⊙

Fortuna major.

○ ○
○ ○
○
○

♀

1. Long life.
2. Abundance.
3. There will be some success.
4. The end will be as desired.
5. Girl.
6. Good and useful.
7. This time he will escape.
8. He will not get his darling.
9. Persons of high positions will divide it with you.
10. The judgment will be in your favour.
11. Success in view.
12. Quietly.
13. No.
14. Fortunate but slow.
15. News speaking of persons of high rank.
16. Take courage.

☉
Fortuna minor.

☿
Conjunctio.

Amissio.
♀

1. Average and changeful.
2. Fortune by trade.
3. Will be abandoned.
4. Tolerable.
5. Twins.
6. They are of the average kind.
7. Dies.
8. He must have her.
9. The legacy will be lost.
10. You will want a settlement.
11. Your fortune is in trading.
12. In bed.
13. The messenger is entering the house.
14. Fortunate.
15. Good.
16. Your adversary is very fortunate.

☿
Conjunctio.

O
O O
O
O O

Amissio.
♀

☉
Fortuna minor.

1. Considerably long but eventful life.
2. Riches by obtaining honourable positions.
3. It will turn out well.
4. Good.
5. Son.
6. They are excessively vain.
7. Recovers.
8. Certain success.
9. After a dispute.
10. The adversary offers a settlement.
11. Advantage.
12. Hot fever.
13. Slow messages.
14. Average.
15. Agreeable.
16. Take courage and go ahead.

Fortuna major.

Populus.

Fortuna major.

⊙

1. Long happy life.
2. Abundance by things connected with water.
3. Fulfilment of wish.
4. As desired.
5. Girl.
6. Many and good servants.
7. Dies.
8. There is nothing to prevent him.
9. Another person has the advantage.
10. Loses.
11. Fortunate on the water or with things connected with it.
12. Beware of danger by water
13. The messenger will arrive soon.
14. Fortunate voyage by water.
15. Average.
16. Go out of the way.

Populus.

Fortuna major.

Fortuna major.

⊙

1. Long life and contented.
2. The possessions are excellent.
3. Go ahead with it.
4. The end will be very good.
5. Girl.
6. You may hire them.
7. Recovers.
8. The other admirers will have to take a back seat.
9. The inheritance is sure.
10. You will gain.
11. You will find what you are seeking.
12. Dies peacefully.
13. Slowly.
14. The voyage on land will be fortunate.
15. Jolly.
16. Attack him.

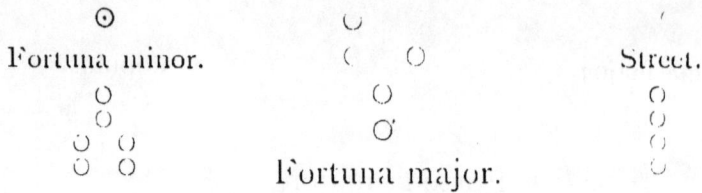

Fortuna minor.   Fortuna major.   Street.

1. Short life, full of disappointments.	9. No.
2. Very little fortune.	10. The adversary succeeds.
3. Disappointment.	11. Disfavour.
4. Only tolerable.	12. Dropsy.
5. Girl.	13. Soon.
6. Faithful.	14. Tolerable.
7. Recovers.	15. Tolerable good news.
8. It is very doubtful.	16. The enemy is stronger than you.

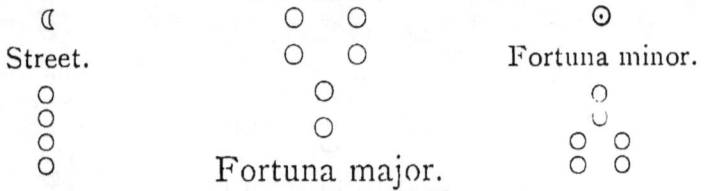

Street.   Fortuna major.   Fortuna minor.

1. Average life.	9. Inheritance.
2. Poverty.	10. Favourable.
3. Partly.	11. Some danger is connected with it.
4. The end is better than the beginning.	12. Fever.
5. Daughter.	13. They are coming.
6. Very vain.	14. Fortunate.
7. Recovers.	15. Good news.
8. She will be his.	16. You will be victorious.

♃
Acquisitio.

○ ○
○ ○

☿
Conjunctio.

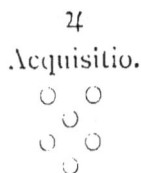

Fortuna major.

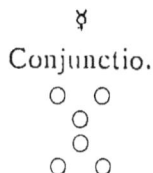

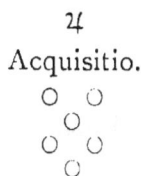

⊙

1. Long life.
2. Average fortune.
3. Yes.
4. Average success.
5. Son.
6. Honest and useful.
7. Dies.
8. He will get what he wants.
9. The profit will not be very great.

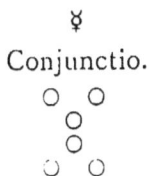

10. Settlement with loss.
11. Some success in mercantile positions.
12. Natural.
13. Soon.
14. Tolerable.
15. Tolerably good.
16. Seek a settlement.

☿
Conjunctio.

○ ○
○ ○

♃
Acquisitio.

Fortuna major.

⊙

1. Long life.
2. Sufficient money.
3. Success.
4. The end is good.
5. Son.
6. You may hire the servant.
7. He will die at last in spite of all hope.
8. The marriage will take place.
9. Unexpected inheritance.

10. You will gain by settling the matter.
11. Fortunate with lawyers.
12. Painless and peaceful.
13. The messenger is on the way.
14. Very good ending.
15. As desired.
16. Your enemy offers a settlement.

♃
Laetitia.

♀
Puella.

Fortuna major.

☉

1. Fortunate and healthy.
2. Riches acquired by marriage.
3. It it is a love affair, it will succeed.
4. Pleasant.
5. Son.
6. Gay, but honest.
7. Will regain his health.
8. He will get his wish.

9. Fortunate.
10. The adversary gains.
11. Fortune through ladies.
12. Peacefully.
13. The messenger will arrive soon.
14. A gay and pleasant voyage.
15. They concern love matters.
16. The enemy is wise, just, and powerful.

♀
Puella.

♃
Laetitia.

Fortuna major.

☉

1. Long and healthy.
2. Abundance.
3. Success.
4. As desired.
5. Son.
6. Good.
7. He will leave his couch.
8. She loves him very affectionately.

9. Inheritance and fortune.
10. A good end.
11. Fortunate with the clergy.
12. Dies in his bed.
13. The letters are detained.
14. Quick and successful.
15. Only good news.
16. The enemy is timorous and unjust.

♂
Rubeus.

☉ ()
() ()

☊
Caput Draconis.

.) ⌣
(`)
(᾿
(᾿

)

U U
U
U
U

Fortuna major.

⊙

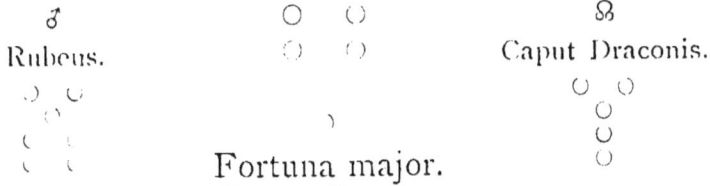

1. Short and sickly life.
2. What has been gained in youth will be lost in old age.
3. Some success.
4. Good.
5. Daughter.
6. Intelligent but dishonest.
7. Recovery.
8. Enjoyment.
9. Several legacies.
10. Favourable judgment.
11. Fortunate among ladies.
12. Natural death.
13. Slowly.
14. Fortunate.
15. Good.
16. You will get the best of your enemy.

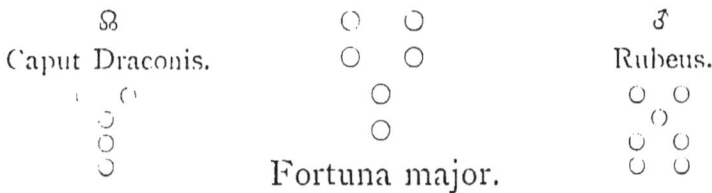

☊
Caput Draconis.

() ()
() ()
()
()

♂
Rubeus.

᾿ (᾿
⌣
()
()

U U
()
U U
U U

Fortuna major.

⊙

1. Short life, full of trouble.
2. Poverty in his youth, rich in old age.
3. The calculation was made without consulting the host.
4. The end will be the ruin of the whole thing.
5. Daughter.
6. Thievish.
7. Death will release him.
8. You are in danger of death on account of your sweetheart.
9. You may inherit a lawsuit.
10. The judge favours your adversary.
11. ♂ is against you.
12. Danger of forcible death.
13. Soon.
14. Dangerous.
15. The news are useless.
16. Your enemy seeks to destroy you,

♂
Puer.

○ ○
○ ○

♉
Cauda Draconis.

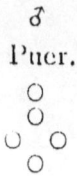

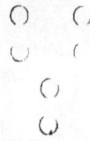

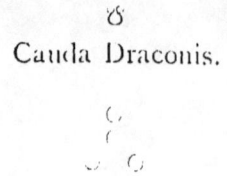

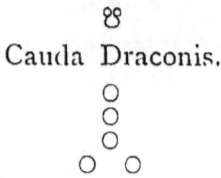

Fortuna major.

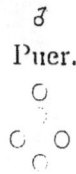

☉

1. Short and painful life.
2. Poverty and misery.
3. It will be a miscarriage.
4. Useless.
5. Son.
6. Lazy, idle, dishonest, thievish.
7. Let him make his preparations for death.
8. It will be useless to think of marriage.
9. The lawyers will divide it among themselves.
10. The lawsuit will be lost.
11. Unfortunate everywhere except in the war.
12. Death by water, drowning or dropsy, etc.
13. No letters will come.
14. A great deal of misfortune.
15. Nothing agreeable.
16. The enemy is too powerful for you.

☊
Cauda Draconis.

○ ○
○ ○

♂
Puer.

Fortuna major.

☉

1. Nothing but misfortune. A short life.
2. What he has will be taken from him by force.
3. Castles in the air.
4. The beginning bad; the end somewhat better.
5. Son.
6. Dishonest, useless.
7. He is on his dying bed.
8. There is many a slip between the cup and the lip.
9. It will not be worth the while to possess it.
10. The adversary loses.
11. There is nothing to be found.
12. Force, executioner, murder, or suicide.
13. You are waiting in vain.
14. You will do well to remain at home.
15. Severe disappointment.
16. You are superior to your enemy.

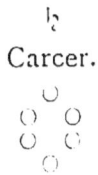

Carcer.

Fortuna major.

Amissio.

⊙

1. Long but sickly life.
2. Inconstant fortune.
3. Useless thoughts about love.
4. Good ending.
5. Daughter.
6. Idle and dishonest.
7. Recovery.
8. The engagement will be broken up.
9. Vain hopes.
10. Neither one of the opponents will gain.
11. Cease to think of it.
12. In bed.
13. They will arrive at last.
14. It will be slow.
15. Trifles.
16. Doubtful ending.

———————

Amissio.

Fortuna major.

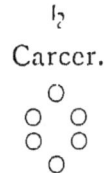

Carcer.

⊙

1. Short life, but a strong constitution.
2. Riches by agriculture or mining.
3. It will slowly succeed.
4. Tolerable.
5. Girl.
6. Average.
7. Dies.
8. There will be a wedding.
9. It amounts to very little.
10. The prospect is bad.
11. Go to the country.
12. Natural death.
13. The mail will not arrive.
14. Tolerable.
15. They are neither good nor bad.
16. The advantage is on your side.

♄
Tristitia.

○ ○
○ ○
○ ○
○

○ ○
○ ○
○ ○
○

☿
Albus.

Fortuna major.

☉

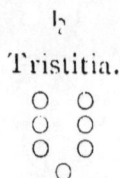

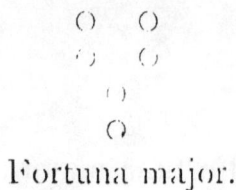

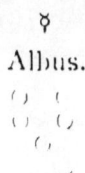

1. Long and healthy life.
2. Riches through writing.
3. Success.
4. The desired end.
5. Daughter.
6. Honest.
7. Will get well.
8. The marriage bells will ring.
9. Good luck.
10. Gains.
11. Fortunate in the legislation.
12. Natural death.
13. Good letters that will come quickly.
14. Fortunate and rapid.
15. Contents are good.
16. Victory.

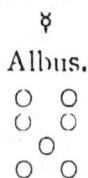

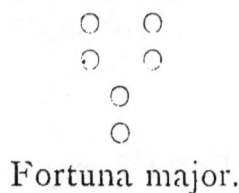

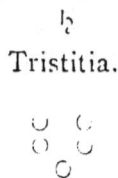

☿
Albus.

○ ○
○ ○
○
○ ○

○ ○
○ ○
○
○

♄
Tristitia.

Fortuna major.

☉

1. Long, melancholy life.
2. Great fortune by agriculture or mining.
3. Success, but slow.
4. Tolerably well.
5. Daughter.
6. Idle, but honest.
7. Dies.
8. Successful in the end.
9. Tolerably lucky.
10. He loses on account of his own neglect.
11. The country is more suitable for you than the city.
12. A natural death.
13. Slowly.
14. Good on the average.
15. Not worth much.
16. The enemy keeps the upper hand.

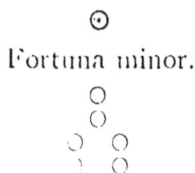

Fortuna minor.

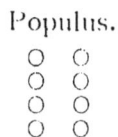

Populus.

Fortuna minor.

1. Long life, considerably fortunate.
2. Some means by trading on the ocean.
3. It has no power to live.
4. Average.
5. Daughter.
6. They will do well enough.
7. Dies.
8. He will be disappointed.
9. Doubtful.
10. Loses.
11. Your fortune is on the water.
12. Painless.
13. The letters will come.
14. Quick and fortunate.
15. Tolerable.
16. The enemy will do no damage.

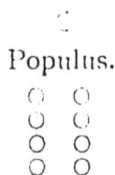

Populus.

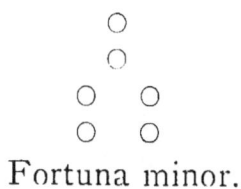

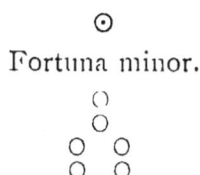

Fortuna minor.

Fortuna minor.

1. Average duration. Changeable.
2. Considerable riches.
3. It will succeed.
4. The end is good.
5. Son.
6. Not the very best.
7. Dies.
8. The wedding will take place.
9. A good legacy will be obtained.
10. Gains.
11. Changeable.
12. Natural death.
13. No.
14. Fortunate and rapid.
15. Good news.
16. Beware of too many enemies.

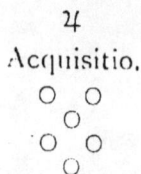
24
Acquisitio.

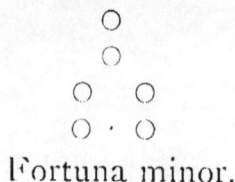

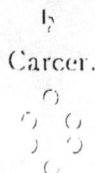
♄
Carcer.

Fortuna minor.

1. Long and healthy life.
2. Rich by works belonging to ♄.
3. No.
4. Average.
5. Son.
6. Honest and industrious.
7. Dies.
8. It is useless to attempt it.
9. Vain hopes.
10. Loses by not coming to time.
11. Ill luck.
12. In bed.
13. No letters.
14. It will be slow.
15. Good on the average.
16. Stay away.

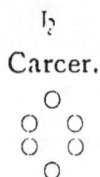
♄
Carcer.

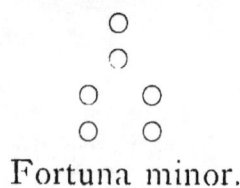

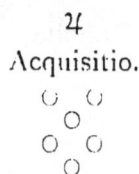
24
Acquisitio.

Fortuna minor.

1. Long life and health.
2. Rich by practising law.
3. It will take place.
4. A good ending.
5. Daughter.
6. They are good.
7. The tomb is waiting for him.
8. Marriage.
9. There is cause for hope.
10. Gains.
11. Fortunate with lawyers.
12. A natural death.
13. They will not arrive.
14. Fortunate.
15. Doubtful.
16. The enemy is harmless.

♃
Laetitia.

♂
Rubeus.

Fortuna minor.

⊙

1. Short, unhealthy life.
2. Loss of property.
3. It will go the wrong way.
4. Unfortunate.
5. Son.
6. Thievish.
7. Dies.
8. Marriage.
9. Worthless.
10. Loss.
11. Fortunate in the army.
12. Forcible death.
13. The letters will be stolen.
14. Unfortunate.
15. Disagreeable.
16. The enemy will come out first best.

♂
Rubeus.

♃
Laetitia.

Fortuna minor.

⊙

1. Long and healthy.
2. Rich from rents.
3. It will take place.
4. The end is good.
5. Son.
6. Good.
7. Recovery.
8. You will be left in the cold.
9. A good legacy.
10. Gains.
11. Fortunate among the clergy.
12. Dies a natural death.
13. Soon.
14. Fortunate.
15. Good news.
16. Your enemy will give way.

Tristitia.

Fortuna minor.

Puer.

1. Short and painful.
2. He will have nothing whatever.
3. His attempts are useless.
4. The end is of no use.
5. Son.
6. Dishonest.
7. Dies.
8. He will marry the woman.
9. Not worth the while to take it.
10. Loses.
11. The stars are against it.
12. He will be killed in battle.
13. Soon.
14. Dangerous voyage.
15. Bad news.
16. The enemy has the advantage.

Puer.

Fortuna minor.

Tristitia.

1. Long and very unfortunate life.
2. A little, obtained with great labour.
3. Extremely doubtful.
4. Doubtful.
5. Daughter.
6. They are lazy.
7. He must die.
8. There will be no marriage, because he will be too careless.
9. A good legacy.
10. Will win at last.
11. Expect nothing.
12. Natural death.
13. No.
14. Delay.
15. Mournful news.
16. The enemy must succumb.

℞   ( )   ♀

Caput Draconis.   ( )   Puella.

      `       `)   ( )

      ( )   ( )   ( )   ( )

Fortuna minor.   ( )

      ( )   ( )

⊙

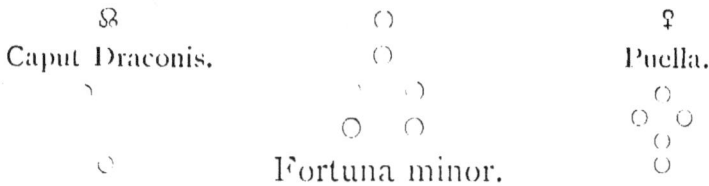

1. Average duration of life. Good circumstances.
2. A fortune through marriage.
3. Fulfilment of wish.
4. A good end.
5. Son.
6. Faithful.
7. Will recover his health.
8. The beloved one will be obtained.
9. Inheritance from ladies.
10. Gains.
11. There is some hope.
12. A natural death.
13. The letters will arrive.
14. Gay and happy voyage.
15. Agreeable news.
16. A settlement is advisable.

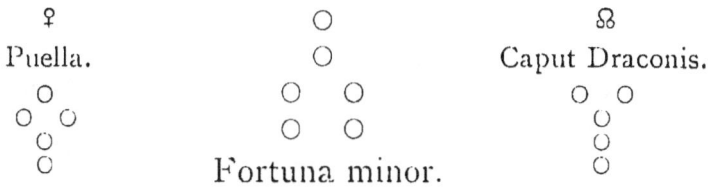

———

♀   O   ℞

Puella.   O   Caput Draconis.

      O   O   O   O   O

    O   O   O   O   O

      O   Fortuna minor.   O

      O   O

⊙

1. Pleasant life but not very long.
2. Good circumstances.
3. Some is true and some of it false.
4. Slow.
5. Daughter.
6. They are good.
7. He will die.
8. He will have no success.
9. Obtains the legacy.
10. It will be settled.
11. Fortune smiles upon you.
12. Natural death.
13. Slowly.
14. Slow but fortunate voyage.
15. Good news.
16. The whole animosity is merely pastime.

♉   ○   ☿
Cauda Draconis   ○   Albus.
        ○   ○ ○   ○
        ○   ○ ○   ○ ○
        ○   ○
    ○ ○   Fortuna minor.   ○ ○

                      ☉

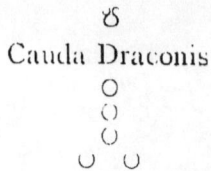

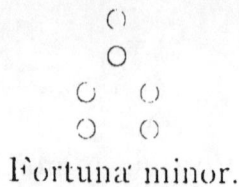

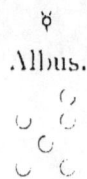

1. Long but unhealthy life.   10. Gains.
2. Rich by trade.   11. Some advantage is waiting
3. Success.   for you.
4. Good ending.   12. A natural death.
5. Daughter.   13. Very soon.
6. Dies.   14. Fortunate.
7. Good.   15. Good news.
8. Never in this case.   16. You will be the master.
9. A good legacy.

☿   ○   ♋
Albus.   ○   Cauda Draconis.
    ○ ○   ○ ○   ○
    ○ ○   ○ ○   ○
        ○   ○
    ○ ○   Fortuna minor.   ○ ○

                      ☉

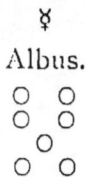

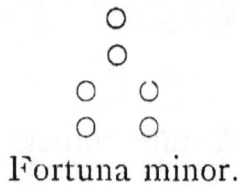

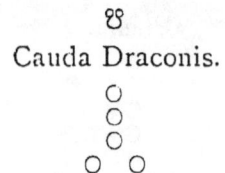

1. Short but healthy life.   10. Is already lost.
2. Poverty and misery.   11. Ill luck.
3. Wrongly calculated.   12. Will be killed.
4. Worthless.   13. Nothing comes.
5. Boy.   14. Unfortunate voyage.
5. They are useless.   15. Without any value what-
7. Dies.   ever.
8. All his efforts are in vain. 16. The enemy is too smart
9. He will get nothing.   for you.

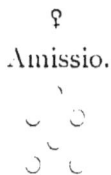
Amissio.

Fortuna minor.

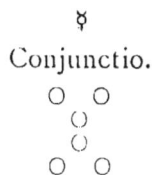
Conjunctio.

⊙

1. Average life. Unhealthy.
2. Average means.
3. Some success.
4. Average.
5. Twins.
6. They are of the average blood.
7. Dies.   [place.
8. The wedding will take

9. It is not very considerable.
10. It will be settled.
11. Fortunate in trading.
12. A natural death.
13. Soon.
14. Good on the average.
15. Neither good nor bad.
16. No harm will be done.

Conjunctio.

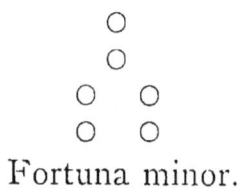
Fortuna minor.

Amissio.

⊙

1. Short and feeble life.
2. Poverty.
3. All in vain.
4. Worthless.
5. Son.
6. They are worthless.
7. Recovery.
8. The marriage will be celebrated.

9. It is of no value.
10. Loss.
11. Bad luck.
12. A natural death.
13. The messenger does not arrive.
14. Unfortunate.
15. Worthless news.
16. A strong enemy.

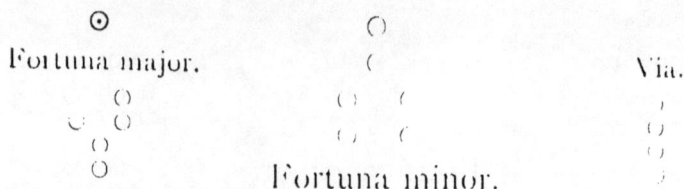

Fortuna major.   Via.

Fortuna minor.

1. Short life, but very fortu-   9. It is worthless.
2. Some fortune.   [nate.   10. Lost.
3. Will be abandoned.   11. Unfortunate.
4. Doubtful.   12. A natural death.
5. Daughter.   13. Soon.
6. Average.   14. Quick and happy.
7. Must die.   15. Of the average kind.
8. He will not get her.   16. Avoid him.

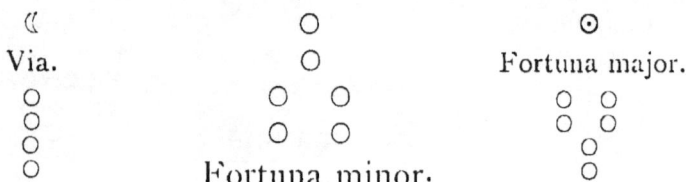

Via.   Fortuna major.

Fortuna minor.

1. Long and healthy life.   9. Fortunate.
2. Riches.   10. Gains.
3. Fulfilment.   11. Very fortunate this time.
4. A good end.   12. Natural.
5. Daughter.   13. Slow.
6. Faithful.   14. Fortunate.
7. Recovery.   15. Good news.
8. Success.   16. You will conquer.

Populus.

Populus.

Populus

Populus

1. Long life.
2. Abundant means.
3. Your dreams will come to nonght.
4. The end is doubtful.
5. Daughter.
6. Honest.
7. Dies.
8. A wedding.
9. You will be disappointed.

10. Thinks look very bad for you.
11. You will be fooled.
12. Natural.
13. Soon.
14. Fortunate and healthy.
15. They are of the average kind.
16. Both adversaries have equal chances.

♃
Acquisitio.

♃
Acquisitio.

Populus,

1. Long and happy.
2. Abundance.
3. You will give up the idea.
4. Average success.
5. Girl.
6. Faithful.
7. Dies.
8. His wish will be granted.

9. A good inheritance.
10. Doubtful.
11. Fortunate.
12. Natural.
13. Delay.
14. Slow but fortunate.
15. Good.
16. You will do him no harm.

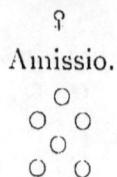

♀
Amissio.

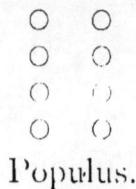

Populus.

☾

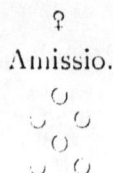

♀
Amissio.

1. Very feeble health.
2. Poverty and misery.
3. The plans will be destroyed.
4. Worthless.
5. Girl.
6. They are useless.
7. Dies.
8. No prospect whatever.
9. Vain hope.
10. Lost.
11. He will obtain nothing.
12. Will be killed.
13. Soon.
14. Unfortunate.
15. Bad news.
16. Loss on both sides.

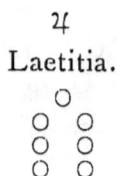

♃
Laetitia.

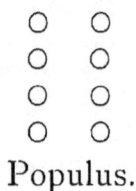

Populus.

☾

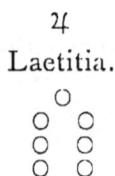

♃
Laetitia.

1. Long and healthy life.
2. Riches.
3. The idea is very good.
4. A good end.
5. Son.
6. They are good.
7. Recovery.
8. She will be his.
9. A fortune by inheritance.
10. Settlement.
11. Luck.
12. Painless and peaceful.
13. Soon.
14. Fortunate.
15. Good.
16. Settlement.

♄

Tristitia.

Populus.

♄

Tristitia.

1. Long and troubled.
2. A little by hard work.
3. Missed.
4. Painful.
5. Girl.
6. Useless.
7. Dies.
8. He will not get her.
9. Worthless.
10. Lost.
11. Nothing to be expected.
12. Natural.
13. No.
14. Slow voyage.
15. Useless.
16. A tricky enemy.

☊

Caput Draconis.

Populus.

☊

Caput Draconis.

1. Average healthy life.
2. Rich by service.
3. It will work, but slowly.
4. Good.
5. Girl.
6. Good.
7. Recovers.
8. He will get her in spite of everything.
9. A good inheritance.
10. Settlement.
11. Lucky.
12. Natural.
13. They are coming.
14. Unfortunate.
15. Good news.
16. Neither one harms the other.

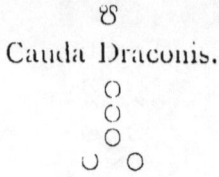

Cauda Draconis.

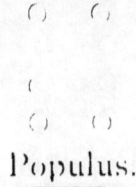

Populus.

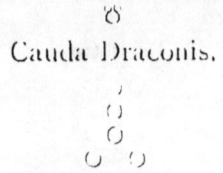

Cauda Draconis.

1. Weak and feeble.
2. Poverty, misfortune.
3. Useless.
4. Unfortunate.
5. Son.
6. Dishonest.
7. Dies.
8. Great disappointment.
9. He will get nothing.
10. Lost.
11. Ill-luck.
12. Will be killed.
13. No.
14. Unfortunate.
15. Worthless.
16. The enemies are a worthless set.

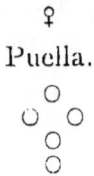

Puella.

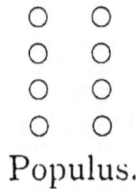

Populus.

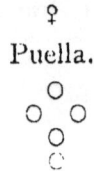

Puella.

1. Long and happy.
2. Rich.
3. Fulfilment.
4. Good.
5. A son.
6. Good.
7. Recovery.
8. He will get her.
9. Yes.
10. Settlement.
11. Unlucky.
12. Natural.
13. They will arrive.
14. Gay voyage.
15. Joyful news.
16. He will do you no harm.

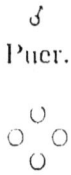

Puer.

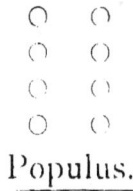

Populus.

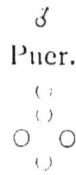

Puer.

1. Wealth and feeble.
2. Poverty in consequence of theft.
3. Non-success.
4. Useless.
5. Son.
6. Worthless.
7. Dies.
8. He will be refused.
9. Gets nothing.
10. The judge is partial.
11. Fortune in war.
12. Forcible.
13. No letters.
14. Quick.
15. Useless.
16. The enemy is dangerous.

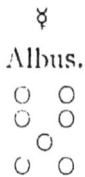

Albus.

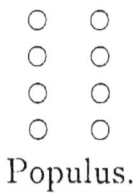

Populus.

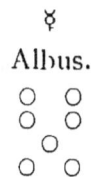

Albus.

1. Long and happy.
2. Riches by service.
3. It will take place.
4. Good.
5. Daughter.
6. Good.
7. Dies.
8. He will succeed.
9. A good inheritance,
10. Settlement.
11. His pen will recommend him.
12. Natural.
13. Soon.
14. Happy.
15. Good.
16. Harmless.

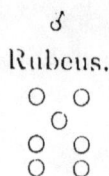

♂
Rubeus.

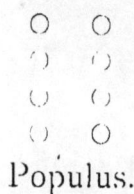

Populus.

☾

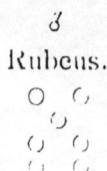

♂
Rubeus.

1. Short and unhealthy.
2. Loss by theft.
3. A wrong beginning.
4. A bad end.
5. Son.
6. Dishonest.
7. Dies.
8. He will be dismissed.
9. Nothing but law-suits.
10. The judge is against you.
11. No.
12. Dies in the war.
13. No letters.
14. Unfortunate.
15. Worthless.
16. A bloodthirsty enemy.

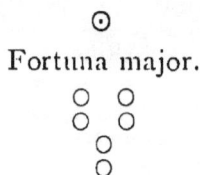

☉
Fortuna major.

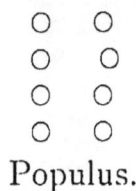

Populus.

☾

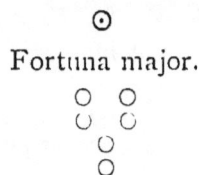

☉
Fortuna major.

1. Fortunate and long.
2. Abundance.
3. It will go on.
4. Fortunate.
5. Girl.
6. Good.
7. Recovery.
8. He will surely get her.
9. A rich legacy.
10. It is already gained.
11. Fortunate among poten-
    tates.
12. Natural.
13. Slowly.
14. Unfortunate.
15. Good.
16. The enemy will give way.

Fortuna minor.

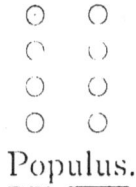

Populus.

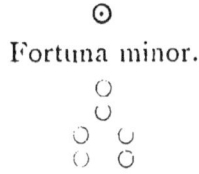

Fortuna minor.

1. Average. Changeable.
2. Satisfactorily.
3. Incorrect.
4. Good.
5. Son.
6. Good.
7. He will gain strength.
8. He will be very fortunate.
9. Gained by a great deal of dispute.
10. A bad prospect.
11. Instability.
12. Natural.
13. Soon.
14. Quick and fortunate.
15. Not very good.
16. You will do him no harm.

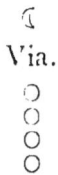

Via.

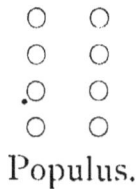

Populus.

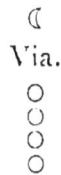

Via.

1. Long and happy.
2. Fortunate in the silver trade.
3. Worthless plans.
4. Tolerable.
5. Daughter.
6. Tolerable.
7. Dies.
8. He is going to marry her.
9. An average legacy.
10. Tolerable success.
11. No.
12. Dies in the water.
13. Soon.
14. Quick and happy.
15. Average.
16. You will get the best of him.

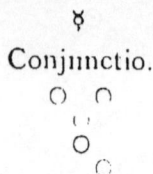

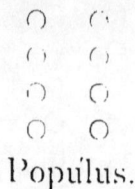

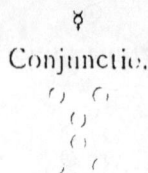

☿

Conjunctio.

☿

Conjunctio.

Populus.

☾

1. Average length.
2. Tolerable.
3. Will turn out truly.
4. Good.
5. Son.
6. Good.
7. Dies.
8. He will obtain her. [nent.
9. Settlement with the oppo-
10. Settlement.
11. Fortunate in the mercantile business.
12. Natural.
13. The letters will arrive.
14. Tolerable.
15. Tolerably good.
16. Settlement.

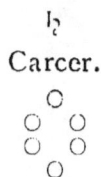

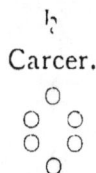

♄

Carcer.

♄

Carcer.

Populus.

☾

1. Long and full of trouble.
2. Riches by agriculture.
3. You will give up the idea.
4. Tolerable.
5. Daughter.
6. They work hard but stupidly.
7. Dies.
8. No prospects,
9. An average inheritance.
10. Unfortunate look-out.
11. No luck.
12. Natural.
13. Slowly.
14. Fortunate.
15. Tolerably good.
16. The enemy will not harm you.

𝄐
Via.
O
O
U
O

O
O
O
O
Via.
𝄐

𝄐
Populus.
O O
O O
O O
O O

1. Long life.
2. Sufficient means.
3. In vain.
4. Tolerable.
5. Daughter.
6. Average.
7. Dies.
8. He is going to obtain her.

9. Without any value.
10. Gained.
11. Changeable luck.
12. In bed.
13. Soon.
14. Quick and fortunate.
15. Tolerably good.
16. You are master.

Populus.
O O
O O
O O
O O

O
O
O
O
Via.
𝄐

Via.
U
U
O
O

1. Short life.
2. Mediocrity.
3. Vain hope.
4. Unfortunate.
5. Girl.
6. Useless.
7. Dies.
8. Yes.
9. Amounts to nothing.

10. Missed the time.
11. Everything is against you.
12. Natural.
13. Soon.
14. Fortunate on land.
15. Trifles.
16. The enemy appears to be master.

♃
Acquisitio.

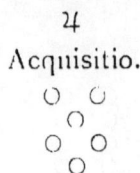

Via.

♌
Amissio.

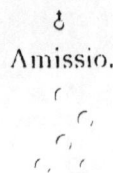

1. Short but healthy.
2. Inconstant.
3. Useless to try it.
4. Deplorable end.
5. Son.
6. Faithful.
7. Recovery.
8. All attempts are in vain.
9. Will do more harm than good.
10. Lost.
11. It is worthless.
12. Natural death.
13. Soon.
14. Quick and fortunate.
15. Disagreeable.   [top.
16. The enemy remains on the

♀
Amissio.

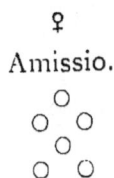

Via.

♃
Acquisitio.

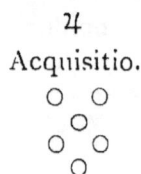

1. Long but not strong.
2. Tolerably good.
3. Good success.
4. Average.
5. Son.
6. Good.
7. He will regain his health.
8. Of course.
9. A good legacy.
10. Gained.
11. Success among lawyers.
12. Peaceful.
13. Slowly.
14. Delay, but fortunate.
15. As desired.
16. The enemy will run away.

4
Laetitia.

Via.

Caput Draconis.

1. Average and healthy.
2. Sufficient.
3. A mistake.
4. Good.
5. Daughter.
6. Faithful.
7. Recovery.
8. He will.
9. It will come to him.
10. Settlement.
11. It is not a very good one.
12. In bed.
13. Slowly.
14. He will not regret it.
15. Pleasant.
16. The enemy becomes a friend.

Caput Draconis.

Via.

Laetitia.

1. Long and joyful.
2. Abundance.
3. Success.
4. Excellent.
5. Son.
6. Good.
7. Recovery.
8. He will obtain his heart's desire.
9. A good legacy.
10. He will be fortunate.
11. Fortunate with the clergy.
12. Natural.
13. They are coming.
14. Fortunate.
15. Pleasant.
16. You will gain the upper hand.

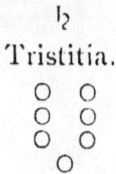

♄

Tristitia.

Via.

☾

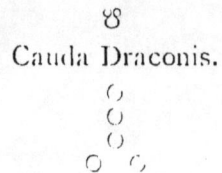

☊

Cauda Draconis.

1. Short and painful.
2. Very little.
3. Non-success.
4. Misfortune.
5. Girl.
6. Treacherous and lazy.
7. Dies.
8. His sweetheart cheats him.
9. The inheritance is dan-
10. Loses the lawsuit. [gerous.
11. No luck.
12. Forcible.
13. Soon.
14. Not very good.
15. Bad news.
16. A strong enemy.

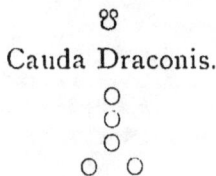

☊

Cauda Draconis.

Via.

☾

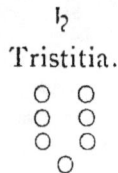

♄

Tristitia.

1. Miserable, but long.
2. A little by hard work.
3. Carrying water in a sieve.
4. A bad end.
5. Girl.
6. Idle.
7. Dies.
8. She will refuse his offer.
9. There is a little to be in-
10. Gains. [herited.
11 More luck on the land than on water.
12. Natural.
13. No.
14. Delay.
15. Mournful news.
16. The enemy loses.

♀
Puella.

○
○
○
○
Via.

☾

♂
Rubeus.

1. Short.
2. Bad.
3. Disappointment.
4. Entirely worthless.
5. Son.
6. Dishonest.
7. He will give up the ghost.
8. You may go ahead.
9. The inheritance escapes.
10. Loses.
11. High places are unhealthy for you.
12. Fever.
13. No letters.
14. Dangerous.
15. Bad.
16. The enemy has the advantage.

———— ——

♂
Rubeus.

○
○
○
○
Via.

☾

♀
Puella.

1. Average.
2. Tolerable.
3. Good.
4. Very well.
5. Son.
6. Good.
7. Recovers.
8. You will soon rejoice.
9. Yes.
10. Gains.
11. Luck.
12. Natural.
13. Soon.
14. Happy.
15. Good.
16. You are superior to him.

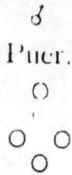

Puer.

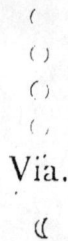

Albus.

Via.

((

1. Long but feeble.
2. Tolerable.
3. Failure.
4. Pleasant.
5. Girl.
6. Good.
7. Dies.
8. The marriage will take place.
9. Yes.
10. Gains.
11. A good prospect.
12. Natural.
13. Certainly.
14. Fortunate.
15. Good.
16. You will make him squirm.

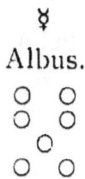

Albus.

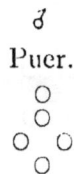

Puer.

Via.

((

1. Short.
2. Bad.
3. Failure.
4. Indifferent.
5. Boy.
6. Dishonest.
7. Dies.
8. Marriage.
9. Bad.
10. Lost.
11. Difficult.
12. Forcible.
13. None.
14. Worthless.
15. Bad.
16. He will conquer you.

Conjunctio.

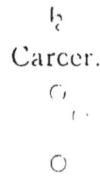

Carcer.

Via.

1. Long.
2. Bad.
3. Failure.
4. Useless.
5. Girl.
6. Idle.
7. Dies.
8. A wedding.

9. Yes.
10. Settlement.
11. No.
12. Natural.
13. Yes.
14. Tolerable.
15. Bad.
16. The enemy succeeds.

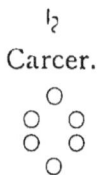

Carcer.

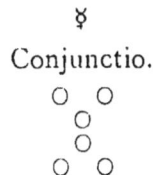

Conjunctio.

Via.

1. Average.
2. Tolerable.
3. It will be accomplished.
4. Indifferent.
5. Girl.
6. Average.
7. He is going to travel.
8. He will be very lucky.
9. It comes in good time.

10. It will be decided in your favour.
11. A tolerably good prospect.
12. Natural.
13. Yes.
14. Tolerably good.
15. Indifferent.
16. The enemy is a loser.

⊙

Fortuna major.

○ ○
○ ○
○
○

○
○
○
○

Via.

(

⊙

Fortuna minor.

(
○
○ (
○ ○

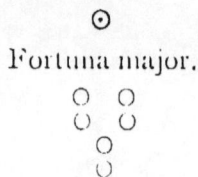

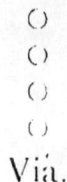

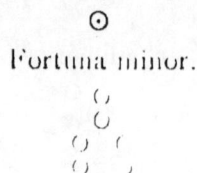

1. Good and long.
2. More than enough.
3. Will disappear.
4. Good.
5. Son.
6. Tolerable.
7. Recovery.
8. She cannot escape him.

9. Good.
10. Lost.
11. Inconstant luck.
12. Natural.
13. They will come.
14. Fortunate.
15. Tolerably good.
16. Victory for the enemy.

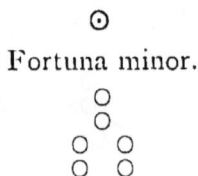

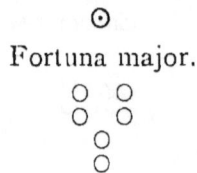

⊙

Fortuna minor.

○
○
○ ○
○ ○

○
○
○
○

Via.

(

⊙

Fortuna major.

○ ○
○ ○
○
○

1. Long but unfortunate.
2. Sufficient.
3. Good.
4. Good.
5. Daughter.
6. Honest.
7. Recovers.
8. Success.

9. A good legacy.
10. Gains.
11. Luck.
12. Natural.
13. Slowly.
14. Delay.
15. Gladness.
16. Victory.

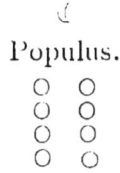

☿
Conjunctio.

○ ○
○
○
○ ○

☾
Populus.

Conjunctio.

☿

1. Long life, indifferent health.
2. Tolerable.
3. False.
4. Indifferent.
5. Daughter.
6. Average.
7. Dies.
8. He will get her.
9. Not much to be had.
10. Loses.
11. No luck.
12. Natural.
13. Soon.
14. Quick.
15. False reports.
16. The adversary desires the peace.

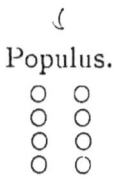

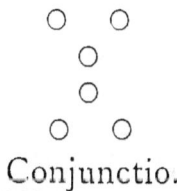

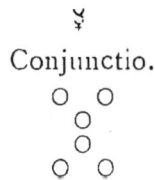

☾
Populus.

○ ○
○
○
○ ○

☿
Conjunctio.

Conjunctio.

☿

1. Average health, short life.
2. A little by trade.
3. A mixed-up affair.
4. Indifferent.
5. Twins.
6. Average.
7. Dies.
8. Marriage.
9. A good legacy.
10. Settlement.
11. A bad position.
12. Natural.
13. Soon.
14. Good.
15. Indifferent.
16. You will want a settlement.

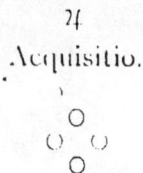

Acquisitio.

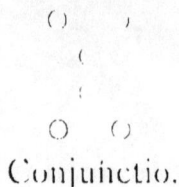

Conjunctio.

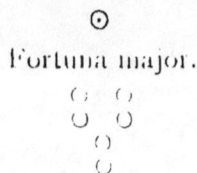

Fortuna major.

1. Long and healthy.
2. Abundance.
3. Success.
4. Glad.
5. Girl.
6. Good.
7. Dies.
8. There is no one to take her from him.
9. Rich.
10. Gains.
11. Luck.
12. Natural.
13. Soon.
14. Fortunate but long.
15. As desired.
16. The adversary will be conquered.

———— ——

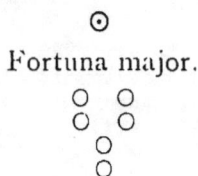

Fortuna major.

Conjunctio.

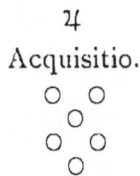

Acquisitio.

1. Healthy and fortunate.
2. Tolerable.
3. Some success.
4. Indifferent.
5. Son.
6. Good enough.
7. Dies.
8. Marriage.
9. Certainly.
10. Gains.
11. Luck among lawyers.
12. Natural.
13. Slowly.
14. Slow.
15. Good news.
16. Enemy remains victorious.

♀

Amissio.

○ ○

Conjunctio.

☿

☉

Fortuna minor.

1. Feeble and short.
2. Little.
3. Deluded hope.
4. Tolerably good.
5. Son.
6. Not the best.
7. Recovers.
8. Obtains his wish.

9. A bad legacy.
10. Inconstant luck.
11. Success.
12. Natural.
13. Soon.
14. In vain.
15. Useless.
16. You will succeed.

——— ———

☉

Fortuna minor.

○ ○

Conjunctio.

☿

♀

Amissio.

1. Feeble and weak.
2. Little constancy.
3. Disappointment.
4. Doubtful.
5. Son.
6. Useless.
7. Dies.
8. Marriage.

9. Cease to think of it.
10. Loses.
11. Nothing.
12. Fever.
13. Soon.
14. Bad.
15. Worthless.
16. No.

♃

Laetitia.

♋ ♋

Cauda Draconis.

Conjunctio.

☿

1. Short but healthy.
2. Bad.
3. All in vain.
4. Useless.
5. Girl.
6. Treacherous.
7. Dies.
8. The engagement will be broken up.
9. The adversary gets it.
10. Lost.
11. Misfortune.
12. Forcible.
13. Soon.
14. Quick.
15. Good.
16. The enemy will be victorious.

♋

Cauda Draconis.

Conjunctio.

☿

♃

Laetitia.

1. Long and happy.
2. Enough.
3. Success.
4. Good.
5. Son.
6. Honest.
7. Recovery.
8. Success.
9. Success.
10. Gains.
11. A good prospect.
12. Natural.
13. They will come.
14. Fortunate.
15. Glad.
16. Enemy will have to leave the field.

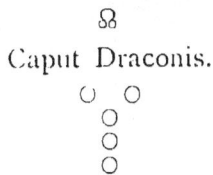

Caput Draconis.

Conjunctio.

Tristitia.

1. Long life.
2. Tolerable.
3. Failure.
4. Indifferent.
5. Daughter.
6. Unfaithful.
7. Dies.
8. Vain hope.
9. Bad prospect.

10. Loses.
11. The stars are against you.
12. Natural.
13. They are coming.
14. Disagreeable.
15. Worthless.
16. Your enemy will be your judge.

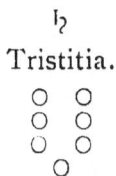

Tristitia.

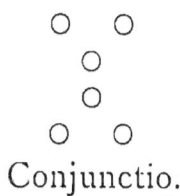

Conjunctio.

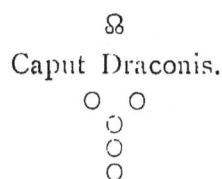

Caput Draconis.

1. Long life.
2. Good circumstances.
3. Fulfilment.
4. Good.
5. Daughter.
6. Faithful.
7. Dies.
8. Success.

9. Success.
10. Success.
11. The fate is in his favour.
12. Natural.
13. They will come.
14. Fortunate.
15. Good.
16. The enemy loses,

☉
Fortuna major.

```
  O  O
  O  O
     O
     O
```

☽ ☾

```
  O   O
```

♃
Acquisitio.

```
        (
      ( )
      ( ) (
```

Conjunctio.

☿

1. Long and happy.
2. Abundance.
3. It will be done.
4. Good.
5. Son.
6. Good.
7. Recovery.
8. Success.
9. Another will get the best.
10. The adversary gains.
11. The lawyers are in your favour.
12. Natural.
13. Slowly.
14. Delay.
15. Good news.
16. Your enemy will be master of the situation.

♃
Acquisitio.

```
  O  O
     O
  O  O
     O
```

```
  O  O
     O
     O
  O  O
```

☉
Fortuna major.

```
  O  O
  O  O
     O
     O
```

Conjunctio.

☿

1. Fortunate and long.
2. Sufficient.
3. Success.
4. Fortunate.
5. Daughter.
6. Honest.
7. Dies.
8. They will become a pair.
9. A good legacy.
10. A good prospect.
11. Favourable.
12. In bed.
13. They will come at last.
14. Postponement.
15. Glad.
16. The adversary gets the worst.

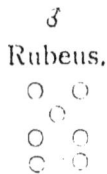

♂
Rubeus.

♀
Albus.

Conjunctio.

☿

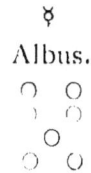

1. Long life.
2. Average riches.
3. Success.
4. Good.
5. Daughter.
6. Faithful.
7. Dies.
8. Marriage.

9. A good legacy.
10. Gains.
11. Favourable.
12. Natural.
13. Soon.
14. Dangerous.
15. Good.
16. The adversary loses.

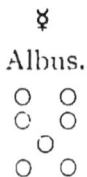

☿
Albus.

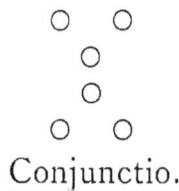

♂
Rubeus.

Conjunctio.

☿

1. Short.
2. Bad.
3. Wrongly calculated.
4. Bad.
5. Son.
6. Useless.
8. Dies.
9. Disappointment.

9. The inheritance will injure [you.
10. Loss.
11. It will be a hell for him.
12. Forcible.
13. None.
14. Slowly.
15. Vexatious.
16. The adversary loses.

☽
Via.

○ ○
○
○
○ ○

♄
Carcer.

Conjunctio.

☿

1. Short life.
2. A little by hard work.
3. From bad to worse.
4. Indifferent.
5. Daughter.
6. Idle.
7. Dies.
8. A wedding.

9. Tolerably good.
10. Gained at last.
11. Secret enemies.
12. Natural.
13. None whatever.
14. Delay.
15. Good.
16. Yes.

♄
Carcer.

○ ○
○
○
○ ○

Via.

Conjunctio.

☿

1. Short.
2. Bad.
3. Disappointment.
4. Indifferent.
5. Daughter.
6. Average.
7. Dies.
8. Marriage.

9. Nothing.
10. Lost.
11. Nothing to be expected.
12. Natural.
13. Certainly.
14. Quick.
15. Indifferent.   [thing.
16. The enemy will gain no-

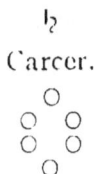

Carcer.

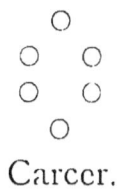

Carcer.

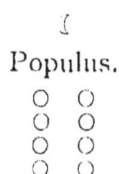

Populus.

1. Long life.
2. Good.
3. It will be interfered with.
4. Average.
5. Daughter.
6. Average.
7. Dies.
8. Undoubtedly.
9. The adversary inherits.
10. Lost.
11. Indifferent luck.
12. Natural.
13. A message.
14. Quick.
15. Indifferent.
16. The enemy must leave the field.

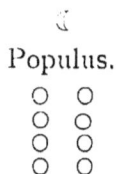

Populus.

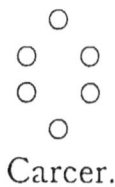

Carcer.

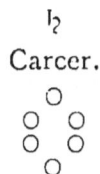

Carcer.

1. Long.
2. Tolerably good fortune in the country.
3. Backward.
4. Tolerable.
5. Daughter.
6. Average.
7. Must go.
8. Cease to think of her.
9. Things look bad.
10. Too late.
11. Nothing to be obtained.
12. Natural.
13. Letters.
14. Delay.
15. Bad.
16. The enemy conquers.

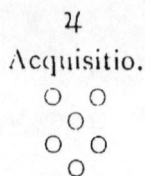

Acquisitio.

Carcer.

♄

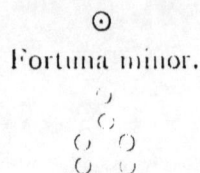

Fortuna minor.

1. Short and happy.
2. Tolerable.
3. In vain.
4. Good.
5. Son.
6. Good.
7. Recovery.
8. Marriage.

9. The legacy is against you.
10. Lost.
11. Lucky, but inconstant
12. Natural.   [fortune.
13. Yes.
14. Soon.
15. Tolerable.
16. You lose.

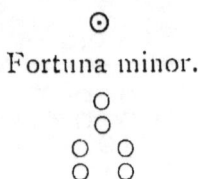

Fortuna minor.

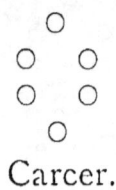

Carcer.

♄

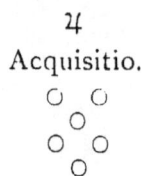

Acquisitio.

1. Long life.
2. Tolerable.
3. It will go ahead.
4. Good.
5. Son.
6. Faithful.
7. Dies.
8. Wedding.

9. A good legacy.
10. Gains.
11. Nothing.
12. Natural.
13. Soon.
14. Fortunate.
15. Good.
16. The enemy returns,

Amissio.

Carcer.

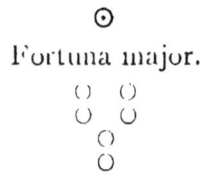

Fortuna major.

1. Long life.
2. Great fortune and durable.
3. It will slowly go on.
4. Happy.
5. Daughter.
6. As good as can be desired.
7. Recovery.
8. He will take the bride.
9. A fat legacy.
10. Gains.
11. A good star.
12. Natural.
13. Letters.
14. Fortunate.
15. Good.
16. Enemy succumbs.

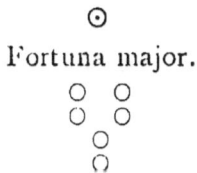

Fortuna major.

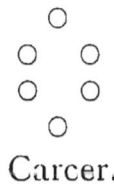

Carcer.

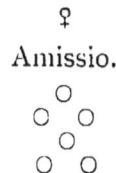

Amissio.

1. Short life.
2. Poverty and misery.
3. The idea is incorrect.
4. Bad.
5. Son.
6. Useless.
7. Must go.
8. ♀ is against you.
9. Another one will rejoice [over it.
10. Gains.
11. An unlucky star.
12. In bed.
13. They have been captured.
14. It will cause you a loss.
15. Contents are vile. [you.
16. Your enemy will step upon

♃
Laetitia.

○
○ ○
○ ○
○ ○

○
○ ○
○ ○
○

Carcer.

♄
Tristitia.

○ ○
○ ○
○ ○
○

♄

1. Long life.
2. Bad.
3. Disappointment.
4. Bad.
5. Daughter.
6. Lazy.
7. Good-bye.
8. At a very late hour.
9. Some success.

10. Decision against you.
11. Everything works against you.
12. Natural.
13. No.
14. Delay.
15. Bad.
16. Enemy comes out first best.

♄
Tristitia.

○ ○
○ ○
○ ○
○

○
○ ○
○ ○
○

Carcer.

♃
Laetitia.

○
○ ○
○ ○
○ ○

♄

1. Long and painful life.
2. Abundant means.
3. Success.
4. Good.
5. Son.
6. Faithful.
7. Recovery.
8. No prospect.

9. Prospect very favourable.
10. Gains.
11. Good luck.
12. Natural.
13. Soon.
14. Fortunate.
15. Glad.
16. Enemy must give way.

♀

Puella.

☽

Albus.

Carcer.

♄

1. Long life.
2. Good.
3. It is well thought out.
4. Good.
5. Daughter.
6. Good.
7. Dies.
8. The stars are against him.
9. It is not very big.

10. Gains.
11. We congratulate.
12. Natural.
13. Soon.
14. Fortunate.
15. Good.
16. The enemy will come out second best.

☿

Albus.

Carcer.

♀

Puella.

♄

1. Short but happy.
2. Tolerable.
3. Good.
4. Indifferent.
5. Son.
6. Faithful.
7. Health.
8. He will get what he loves.

9. There is a little of it.
10. Gains.
11. Good luck.
12. Natural.
13. Soon.
14. Fortunate.
15. Good.
16. The victory is doubtful.

♂

Rubeus.

O O
  O
O O
O O

()
() ()
() ()
()

Carcer.

♄

♂

Puer.

()
O
O O
O

1. Short but unfortunate.
2. Poor and miserable.
3. Incorrect.
4. An arrest and imprisonment.
5. Son.
6. Dishonest.
7. Dies.
8. No success.

9. It is harmful.
10. Loses.
11. Obstacles.
12. Forcible.
13. No letters.
14. Unfortunate.
15. Useless.
16. The enemy will be
    victorious.

——— ——

♂

Puer.

O
O
O O
O

O
O O
O O
O

Carcer.

♄

♂

Rubeus.

O O
 O
O O
O O

1. Long life.
2. Loss of property
3. Let it alone.
4. Misery.
5. Son.
6. Thievish.
7. Dies.
8. The marriage will be
   postponed for ever.

9. Nothing but quarrels.
10. Gains, but will receive no
    advantage from it.
11. Avoid it.
12. Forcible.
13. None.
14. Dangerous.
15. Throw them into the fire.
16. Victory.

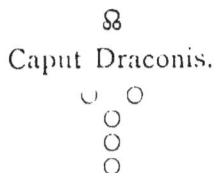

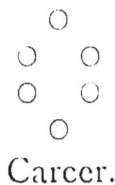

☊ Caput Draconis.

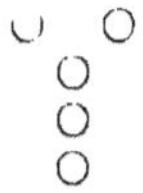

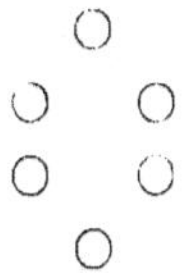

Carcer.

♄

☋ Cauda Draconis.

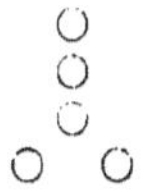

1. Short.
2. Poverty.
3. A false calculation.
4. Bad.
5. Son.
6. Greedy.
7. Dies.
8. No success.
9. Nothing to inherit for you.
10. Loses.
11. The position is dangerous.
12. In prison.
13. None.
14. Unfortunate.
15. Useless.
16. The enemy is too cunning.

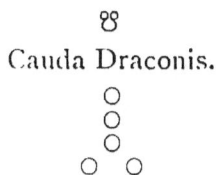

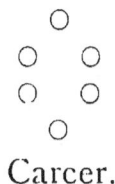

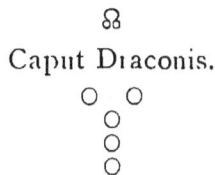

☋ Cauda Draconis.

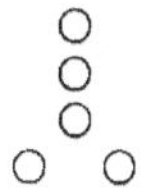

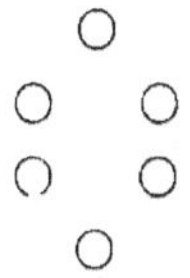

Carcer.

♄

☊ Caput Draconis.

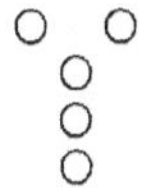

1. Short and unhappy.
2. Average means.
3. Succeeds.
4. Good.
5. Daughter.
6. Good.
7. Dies.
8. No marriage.
9. He will get it,
10. Gains.
11. Luck.
12. Natural.
13. Slowly.
14. Slow.
15. Agreeable.
16. The adversary loses the game,

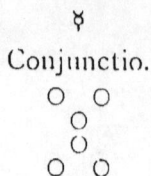

Conjunctio.

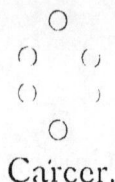

Carcer.

Via.

℞

1. Short and painful.
2. Poor.
3. Disappointment.
4. Indifferent.
5. Daughter.
6. Average.
7. Dies.　　　[discontinued.
8. The engagement will be

9. Bad prospects.
10. Loses.
11. No luck.
12. Natural.
13. A message.
14. Quick.
15. Tolerable.
16. The enemy desires peace.

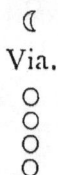

Via.

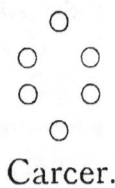

Carcer.

℞

Conjunctio.

1. Average.
2. Tolerable.
3. Bad prospects.
4. Bad.
5. Twins.
6. Average.
7. Dies.
8. He will get her.
9. Pays for the mourning.

10. Decided to your advantage.
11. Everything looks bad.
12. Natural.
13. The messenger is near.
14. Profitable.
15. Good.
16. Victory.

THEOSOPHICAL PUBLICATION SOCIETY,

7, DUKE STREET, ADELPHI,

LONDON, W.C., 1889,

AND

"The Path," P.O. Box 2,659, New York, U.S.A.

The Theosophical Publication Society has been established for the purpose of being a centre from which Theosophical Literature can be disseminated, and made available for all those who wish to learn something respecting those deeper truths of Science, Philosophy and Religion, which are now offered to the world.

For this purpose Pamphlets are published in a cheap form, and are sent post free to every subscriber to the Society. The Pamphlets consist of original essays explanatory of Theosophy, as well as reprints of articles of value at present buried in the back numbers of magazines.

The Society proposes, in addition, to publish works on Theosophy, Occult Science, and kindred subjects, and also to bring out translations of valuable articles and books written in other languages.

The following Pamphlets have already been published:—

No.			Price
1. March.	"Theosophy and the Churches"	.. ..	2d.
2. ,,	"Psychic Bodies" and "Soul Survival"	..	2d.
3. April.	"Philosophie der Mystik"	.. ..	3d.
4. ,,	"The Theosophical Movement," &c. ..	..	2d.
5. May.	"What is Matter and what is Force?" &c.	..	2d.
6. ,,	"Reincarnation," &c., &c.	.. ..	3d.
7. June.	"Practical Occultism" (for members only).		
8. ,,	"Epitome of Theosophical Teachings"	..	2d.
9. July.	"Keely's Secrets"	.. .. ..	6d.
10. August.	"Nature Spirits or Elementals"	.. ..	3d.
11. Sept.	"The Higher Science" ..	.. ..	2d.
12. ,,	"Was Jesus a Perfect Man?" ..	.. ..	2d.
13. Oct.	"The Hebrew Talisman"	.. ..	6d.
14. Nov.	"Selflessness," "Thelyphthoria," and "Taro"		3d.
15. Dec.	"Swedenborg Bifrons" ..	.. ..	3d.
16. ,,	"Theosophical Concepts of Evolution and Religion"	..	3d.

Subscription, 5s. per annum, post free.